T0237577

SpringerBriefs in Statistics

JSS Research Series in Statistics

The current research of statistics in Japan has expanded in several directions in line with recent trends in academic activities in the area of statistics and statistical sciences over the globe. The core of these research activities in statistics in Japan has been the Japan Statistical Society (JSS). This society, the oldest and largest academic organization for statistics in Japan, was founded in 1931 by a handful of pioneer statisticians and economists and now has a history of about 80 years. Many distinguished scholars have been members, including the influential statistician Hirotugu Akaike, who was a past president of JSS, and the notable mathematician Kiyosi Itô, who was an earlier member of the Institute of Statistical Mathematics (ISM), which has been a closely related organization since the establishment of ISM. The society has two academic journals: the Journal of the Japan Statistical Society (English Series) and the Journal of the Japan Statistical Society (Japanese Series). The membership of JSS consists of researchers, teachers, and professional statisticians in many different fields including mathematics, statistics, engineering, medical sciences, government statistics, economics, business, psychology, education, and many other natural, biological, and social sciences. The JSS Series of Statistics aims to publish recent results of current research activities in the areas of statistics and statistical sciences in Japan that otherwise would not be available in English; they are complementary to the two JSS academic journals, both English and Japanese. Because the scope of a research paper in academic journals inevitably has become narrowly focused and condensed in recent years, this series is intended to fill the gap between academic research activities and the form of a single academic paper. The series will be of great interest to a wide audience of researchers, teachers, professional statisticians, and graduate students in many countries who are interested in statistics and statistical sciences, in statistical theory, and in various areas of statistical applications.

More information about this series at http://www.springer.com/series/13497

Masanori Sawa · Masatake Hirao ·
Sanpei Kageyama

Euclidean Design Theory

 Springer

Masanori Sawa
Graduate School of System Informatics
Kobe University
Kobe, Hyogo, Japan

Masatake Hirao
School of Information and Science
Technology
Aichi Prefectural University
Nagakute, Aichi, Japan

Sanpei Kageyama
Research Center for Mathematics
and Science Education
Tokyo University of Science
Tokyo, Japan

Hiroshima University
Hiroshima, Japan

ISSN 2191-544X   ISSN 2191-5458 (electronic)
SpringerBriefs in Statistics
ISSN 2364-0057   ISSN 2364-0065 (electronic)
JSS Research Series in Statistics
ISBN 978-981-13-8074-7   ISBN 978-981-13-8075-4 (eBook)
https://doi.org/10.1007/978-981-13-8075-4

This Springer imprint is published by the registered company Springer Nature Singapore Pte Ltd.
The registered company address is: 152 Beach Road, #21-01/04 Gateway East, Singapore 189721, Singapore

Preface

Research in the area of discrete optimal designs has been steadily and rapidly growing, especially during past several decades. The number of publications available in this area is in several hundreds. The optimality problems have been formulated in various models arising in the experimental designs and substantial progress has been made toward solving these problems. In the meantime, the theory of mainly continuous optimal designs has been reviewed and reorganized in a comprehensive survey by Pukelsheim's book *Optimal Design of Experiments* (1993; John Wiley) of over 400 pages, which also tries to give a unified optimality theory of embracing a wide variety of design problems. Further developments can be found in a book *Topics in Optimal Design* (2002; Springer) by Liski, Mandal, Shah, and Sinha who cover a wide range of topics in both discrete and continuous optimal designs.

We want to deal with the construction of optimal experimental designs pure-mathematically and systematically under a new framework. The aim of the present book is to show the modern first treatment of experimental designs for giving a comprehensive introduction to the interrelationship of the theory of optimal designs with the theory of cubature formulas in numerical analysis. It also provides the reader with original new ideas for constructing optimal designs, though this is not a full-length treatment of the subject.

The book opens with the basics on reproducing kernel Hilbert space, and builds up to more advanced topics including, bounds for the number of cubature formula points, equivalence theorems for statistical optimalities, and the Sobolev theorem for the cubature formula. It ends with a generalization for the abovementioned classical results in a functional analytic manner.

Since papers on optimal designs are published in a variety of journals, and because of the extensive role of these designs in design of experiments and other areas we believe it is imperative to gather these results and present them in varied form to suit diverse interests. This book is an instance of such an attempt.

As the Contents show, the material is covered in five chapters. Chapter 1 provides a brief summary of basic ideas and facts concerning kernel functions which are closely related to the theories of cubature formula in numerical analysis as well as of Euclidean design which is a special point configuration in the Euclidean space.

Chapter 2 is to show that cubature formulas can be used for finding optimal designs of experiment as a statistical application. In this sense, the relationship between cubature formulas and Euclidean designs is discussed. Chapter 3 is devoted to organically combining optimal experimental designs and Euclidean designs in algebraic combinatorics. Chapter 4 discusses and describes two advanced methods of constructing optimal Euclidean designs, one based on orbits of reflection groups and another based on combinatorial or statistical subjects such as combinatorial t-designs and orthogonal arrays. The climax of this book, Chap. 5, introduces the concept of generalized cubature formula and lays the foundation of Euclidean Design Theory, which not only produces a novel framework for understanding optimal designs, based on the theory of cubature formulas in analysis and spherical or Euclidean designs in combinatorics, but also finds some applications to design of experiments.

This book is especially intended for readers who are interested in recent advances in the construction theory of cubature formulas, Euclidean designs, and optimal experimental designs. Moreover, it is recommended to research workers who seek rich interactions between optimal experimental designs and various mathematical subjects including spherical design theory in combinatorics, embedding theory of Banach spaces in functional analysis, and cubature theory in numerical analysis. A novel communicating platform is finally provided for "design theorists" in a wide variety of research fields. Of course, since we are also aiming at an audience with a wide range of backgrounds, including postgraduate students in statistics or combinatorics or both, we have assumed a reasonable knowledge of linear algebra, analysis, finite field theory, but very little else. Number theory and some essential statistical or analytical concepts are developed as needed. If you would glance at titles in the references, you might conceive how to deal with topics on these problems.

It is hoped that the book will also be useful as a secondary and primary reference for statisticians and mathematicians doing research on the design of experiments, and also for the experimenters in diverse fields of applications.

Acknowledgments Masanori Sawa sincerely appreciates his wife Ikumi and three children, Kojiro, Kenji, and Shiori, who have been patiently watching him indulging in writing the present book during holidays or even vacations. He would also like to thank his mother Kazuko, his late father Seiji, and the family of uncle, Koji, Sumiko, Takuji for their warm supports so far. Masatake Hirao is eagerly thankful to his family for their support and encouragement, especially his wife Tomoko and wonderful child Rintaro. Finally, we would like to thank Prof. M. Iwasaki, Yokohama City University, who is one of Series editors, JSS Research Series in Statistics, for his warm encouragements of preparing our draft for the Series.

Kobe, Japan   Masanori Sawa
Nagakute, Japan   Masatake Hirao
Hiroshima, Japan   Sanpei Kageyama
May 2019

Contents

1 Kernel Functions 1
   1.1 Kernels and Polynomials 1
   1.2 Kernels and Compact Formulas 6
   1.3 Kernels and Dimension 11
   1.4 Further Remarks 14
   References 17

2 Cubature Formula 19
   2.1 Cubature Formula and Elementary Construction Methods 20
   2.2 Existence Theorems and Lower Bounds 23
   2.3 Euclidean Design and Spherical Symmetry 31
   2.4 Further Remarks and Open Questions 38
   References 42

3 Optimal Euclidean Design 45
   3.1 Regression and Optimality 45
   3.2 Optimal Euclidean Design and Characterization Theorems 51
   3.3 Realization of the Kiefer Characterization Theorem 55
   3.4 Further Remarks and Open Questions 58
   References 60

4 Constructions of Optimal Euclidean Design 63
   4.1 Finite Reflection Groups 64
      4.1.1 Group A_d 65
      4.1.2 Group B_d 68
      4.1.3 Group D_d 71
   4.2 Invariant Polynomial and the Sobolev Theorem 73
      4.2.1 Group A_d 76
      4.2.2 Group B_d 79
      4.2.3 Group D_d 80

4.3   Corner Vector Methods.   83
    4.3.1   Group A_d   85
    4.3.2   Group B_d   87
    4.3.3   Group D_d   91
4.4   Combinatorial Thinning Methods   94
4.5   Further Remarks and Open Questions   98
References   100

5   **Euclidean Design Theory**   103
5.1   Generalized Cubature Formula   104
5.2   Generalized Tchakaloff Theorem   108
5.3   Generalized Fisher Inequality   112
5.4   Generalized LP Bound   116
5.5   Generalized Sobolev Theorem   119
5.6   Conclusion and Further Implications   121
References   128

Correction to: Euclidean Design Theory C1

Index   131

Chapter 1
Kernel Functions

The present chapter provides a brief summary of basic ideas and facts concerning *kernel functions*, which are closely related to the theories of cubature formula being a certain class of integration formulas in numerical analysis, as well as of Euclidean design, which is a special point configuration in the Euclidean space.

The first three sections review some general elementary facts such as the Aronszajn theorem and the Riesz representation theorem, after which the emphasis gradually shifts to technical details, including explicit computations of kernel functions and dimensions of finite-dimensional normed vector spaces. The final section covers related topics such as the Seidel'nikov inequality in discrete geometry [14] and a novel connection between kernel machines and cubature formulas [5].

The aim of this chapter is to match the breadth of theory of kernel functions and that of cubature formulas. Each section includes results about cubature formulas without detailed explanations on terminologies; such details will be explained in the subsequent chapters. The prerequisite for reading this chapter is a knowledge of functional analysis at the standard undergraduate level, but the reader who has not learned it can still read this chapter by accepting some advanced materials in spherical harmonics and orthogonal polynomials.

Most of the materials concerning kernel functions, which appear in this chapter without proofs, can be found in [13]. For an extensive treatment of spherical harmonics and orthogonal polynomials, we refer the reader to [16, 17].

1.1 Kernels and Polynomials

At first, the definition of kernel function is described. Throughout let Ω be a set, usually, of real vectors, or more generally a topological space.

The original version of this chapter was revised: Missed out author corrections have been incorporated. The correction to this chapter is available at https://doi.org/10.1007/978-981-13-8075-4_6.

© The Author(s), under exclusive license to Springer Nature Singapore Pte Ltd. 2019   1
M. Sawa et al., *Euclidean Design Theory*, JSS Research Series in Statistics,
https://doi.org/10.1007/978-981-13-8075-4_1

Definition 1.1 (*Kernel function*) A function $K : \Omega \times \Omega \to \mathbb{R}$ is called a *kernel* (*function*) *on* Ω if

(i) K is positive semi-definite, namely

$$\sum_{i,j=1}^{n} c_i c_j K(\omega_i, \omega_j) \geq 0 \quad \text{for all } n \geq 1, c_i \in \mathbb{R} \text{ and } \omega_i \in \Omega, \qquad (1.1)$$

(ii) K is symmetric, namely

$$K(\omega, \omega') = K(\omega', \omega) \quad \text{for all } \omega, \omega' \in \Omega.$$

The following fact is fundamental in the theory of kernel functions.

Theorem 1.1 (Aronszajn theorem, [2]) *Let K be a kernel function on a set Ω. Then there exists a unique Hilbert space (say, $\mathscr{H}_K$) with inner product $(\cdot, \cdot)_{\mathscr{H}_K}$ such that*

$$f_\omega := K(\cdot, \omega) \in \mathscr{H}_K \quad \text{for every } \omega \in \Omega,$$
$$f(\omega) = (f(\omega'), K(\omega', \omega))_{\mathscr{H}} \quad \text{for every } \omega \in \Omega \text{ and } f \in \mathscr{H}_K. \qquad (1.2)$$

The latter condition of (1.2) is called the *reproducing property*. Some authors, thus, use the term "reproducing kernels" for kernel functions K and accordingly "reproducing kernel Hilbert spaces (RKHS)" for the corresponding spaces $\mathscr{H}_K$.

Remark 1.1 Throughout the present book, we are mainly concerned with finite-dimensional normed vector spaces of functions, which are always Hilbert spaces. Most of the results given in this chapter are described over the field of real numbers, some of which will also be discussed over the field of complex numbers in Chap. 5.[1] It is also assumed that functional spaces considered in the present book always have a basis.

Let $\{f_i\}_{i=1}^{n}$ be a basis of an n-dimensional real vector space $\mathscr{H}$ of $\mathbb{R}$-functions on a set Ω with inner product $(\cdot, \cdot)_{\mathscr{H}}$. Then any vector $f \in \mathscr{H}$ is expressed in the form of $f = \sum_i c_i f_i$. It follows that

$$\|f\|_{\mathscr{H}} := \sqrt{(c_1, \ldots, c_n) A (c_1, \ldots, c_n)^T}$$

can define a norm of space $\mathscr{H}$, when $A = (a_{ij})$, $a_{ij} = (f_i, f_j)_{\mathscr{H}}$, is a positive-definite matrix, where $(c_1, \ldots, c_n)^T$ is the transpose of $(c_1, \ldots, c_n)$. The space $\mathscr{H}$ equipped with norm $\| \cdot \|_{\mathscr{H}}$ has a unique kernel

$$K(\omega, \omega') = \sum_{i,j=1}^{n} b_{ij} f_i(\omega) f_j(\omega'), \qquad (1.3)$$

---

[1] Some results described in this chapter can also be proved for separable Hilbert spaces.

where $B = (b_{ij})$ is the inverse of A. In particular when $\{f_i\}_{i=1}^n$ is orthonormal,[2] the right-hand side of (1.3) is expressed as follows:

$$\sum_{i,j=1}^n b_{ij} f_i(\omega) f_j(\omega') = \sum_{i=1}^n a_{ii}^{-1} f_i(\omega) f_i(\omega') = \sum_{i=1}^n \frac{1}{\|f_i\|_{\mathscr{H}}^2} f_i(\omega) f_i(\omega')$$

as will be seen in Example 1.1.

Remark 1.2 With the above notation, A is the so-called *Gram matrix*. The *information matrix*, as will be available in Chap. 3, is a class of Gram matrices. Note that

$$\sum_{i,j=1}^n b_{ij} f_i(\omega) f_j(\omega') = (f_1(\omega), \ldots, f_n(\omega)) B (f_1(\omega'), \ldots, f_n(\omega'))^T.$$

Hence for $(f_i, f_j)_{\mathscr{H}} := \int_\Omega f_i(\omega) f_j(\omega) \mu(d\omega)$ for probability measures μ on Ω, (1.3) is related to the famous *G-optimality* in the theory of optimal designs [8]. It is quite likely that most of the previously published works concerning optimal designs have been written without being aware of this viewpoint.

Example 1.1 (Kernel polynomial) Denote by $\mathscr{P}(\mathbb{R})$ the space of all polynomials defined on the real line $\Omega := \mathbb{R}$, with inner product given by

$$(f, g)_{\mathscr{P}(\mathbb{R})} := \frac{1}{\int_{\mathbb{R}} \exp(-u^2) du} \int_{\mathbb{R}} f(u) g(u) \exp(-u^2) \, du.$$

Let $\mathscr{H} := \mathscr{P}_t(\mathbb{R})$ be the subspace of $\mathscr{P}(\mathbb{R})$ that consists of all polynomials of degree at most t and let $\{f_i\}_{i=1}^{t+1}$ be an orthonormal basis. Then the function

$$K(u, u') := \sum_{i=1}^{t+1} \frac{1}{\|f_i\|_{\mathscr{P}(\mathbb{R})}^2} f_i(u) f_i(u'), \quad u, u' \in \mathbb{R} \tag{1.4}$$

is a kernel function on $\mathbb{R}$, where $\|f\|_{\mathscr{P}(\mathbb{R})} := (f_i, f_i)_{\mathscr{P}(\mathbb{R})}^{1/2}$. It is obvious that the function K is symmetric. Condition (1.1) also holds since

$$\sum_{i,j=1}^n c_i c_j K(u_i, u_j) = \sum_{i,j=1}^n c_i c_j \sum_{k=1}^{t+1} \frac{1}{\|f_k\|_{\mathscr{P}(\mathbb{R})}^2} f_k(u_i) f_k(u_j)$$

---

[2]The definition of orthonormal basis will be given in the next section.

$$= \sum_{k=1}^{t+1} \frac{1}{\|f_k\|_{\mathscr{P}(\mathbb{R})}^2} \sum_{i,j=1}^{n} c_i c_j f_k(u_i) f_k(u_j)$$

$$= \sum_{k=1}^{t+1} \left(\frac{1}{\|f_k\|_{\mathscr{P}(\mathbb{R})}} \sum_{i=1}^{n} c_i f_k(u_i) \right)^2$$

$$\geq 0.$$

Kernel (1.4) is closely tied with the theory of quadrature formulas, as will be seen in Theorem 1.3. The notion of kernel polynomials can also be considered for multivariate polynomial spaces, as will be seen in Theorem 2.4 in Sect. 2.2.

Now some elementary properties about kernel functions are reviewed for further arguments below.

Proposition 1.1 *Let a_1, a_2 be nonnegative real numbers and $K^{(1)}, K^{(2)}$ be kernel functions on a set Ω. Then the following hold:*

(i) *$a_1 K^{(1)} + a_2 K^{(2)}$ is a kernel on Ω;*
(ii) *$K^{(1)} K^{(2)}$ is a kernel on Ω.*

Proof Statement (i) is straightforward.

To see (ii), first let $K^{(\ell)}(\omega_i, \omega_j) := K_{ij}^{(\ell)}$ for $\ell = 1, 2$. Then the matrix $(K_{ij}^{(1)})_{i,j}$ is positive semi-definite and so by the singular value decomposition there exist mutually orthogonal row vectors $v_1, \ldots, v_d \in \mathbb{R}^d$ and nonnegative reals $\lambda_1, \ldots, \lambda_d$ such that

$$(K_{ij}^{(1)})_{i,j} = \sum_{k=1}^{d} \lambda_k v_k^T v_k.$$

Hence

$$\sum_{i,j} c_i c_j K_{ij}^{(1)} K_{ij}^{(2)} = \sum_{i,j} c_i c_j \left(\sum_k \lambda_k v_{ki} v_{kj} \right) K_{ij}^{(2)}$$

$$= \sum_k \lambda_k \sum_{i,j} (c_i v_{ki})(c_j v_{kj}) K_{ij}^{(2)}$$

$$\geq 0.$$

Here, the positive semi-definiteness of $K^{(2)}$ is used in the last inequality and also v_{ki} is the kth coordinate of v_i. $\square$

Example 1.2 (Polynomial kernel) For column vectors $\omega, \omega' \in \mathbb{R}^d$, ordinary Euclidean inner product

$$\langle \omega, \omega' \rangle := \omega^T \omega' := \sum_{j=1}^{d} \omega_j \omega'_j$$

is a kernel function on $\mathbb{R}^d$, which is often used in linear discriminant analysis (see, e.g., [4]). More generally, the ℓ-th power $\langle \omega, \omega' \rangle^\ell$ for all ℓ is a kernel function on $\mathbb{R}^d$ by Proposition 1.1 (ii).

The next result characterizes Hilbert spaces that correspond to the sum and product of given kernels.

Theorem 1.2 (e.g., [13]) *Let $\mathcal{H}_\ell$, $\ell = 1, 2$, be the Hilbert space corresponding to a kernel function $K^{(\ell)}$ on a set Ω, together with inner product $(\cdot, \cdot)_{\mathcal{H}_\ell}$. Then the following hold:*

(i) The sum $K := K^{(1)} + K^{(2)}$ corresponds to the Hilbert space

$$\mathcal{H}_K := \{f_1 + f_2 \mid f_1 \in \mathcal{H}_1, \ f_2 \in \mathcal{H}_2\}$$

with norm

$$\|f\|^2_{\mathcal{H}_K} = \min\{\|f^{(1)}\|^2_{\mathcal{H}_1} + \|f^{(2)}\|^2_{\mathcal{H}_2} \mid f = f^{(1)} + f^{(2)}, \ f^{(1)} \in \mathcal{H}_1, f^{(2)} \in \mathcal{H}_2\}.$$

(ii) The product $K^{(1)} K^{(2)}$ corresponds to the Hilbert space

$$\mathcal{H}_K := \left\{ \sum_j f_j^{(1)} f_j^{(2)} \mid f_j^{(1)} \in \mathcal{H}_1, \ f_j^{(2)} \in \mathcal{H}_2, \ \sum_j \|f_j^{(1)}\|^2_{\mathcal{H}_1} \|f_j^{(2)}\|^2_{\mathcal{H}_2} < \infty \right\}$$

with norm

$$\|f\|^2_{\mathcal{H}_K} = \min\left\{ \sum_j \|f_j^{(1)}\|^2_{\mathcal{H}_1} \|f_j^{(2)}\|^2_{\mathcal{H}_2} \mid f = \sum_j f_j^{(1)} f_j^{(2)}, \ f_j^{(1)} \in \mathcal{H}_1, f_j^{(2)} \in \mathcal{H}_2 \right\}.$$

Example 1.3 (Modified kernel) Let $\mathcal{H}_1 := \mathscr{P}^*_{2e-1}(\mathbb{R})$ (resp. $\mathcal{H}_2 := \mathscr{P}^*_{2e}(\mathbb{R})$) be the space of polynomials of odd degree $\leq 2e - 1$ (resp., of polynomials of even degree $\leq 2e$). Let $\mathscr{F}_1 := \{f_i^{(1)}\}_{i=1}^e$ (resp. $\mathscr{F}_2 := \{f_i^{(2)}\}_{i=1}^{e+1}$) be an orthonormal basis of $\mathcal{H}_1$ (resp. $\mathcal{H}_2$) with respect to $(\cdot, \cdot)_{\mathscr{P}(\mathbb{R})}$. An argument similar to the calculation done in Example 1.1 shows that the function

$$K^{(\ell)}(u, u') := \sum_i \frac{1}{\|f_i^{(\ell)}\|^2_{\mathscr{P}(\mathbb{R})}} f_i^{(\ell)}(u) f_i^{(\ell)}(u'), \quad u, u' \in \mathbb{R}$$

is a kernel for $\mathcal{H}_\ell$, respectively. Some authors call them *modified (polynomial) kernels* [6]. By Theorem 1.2 (i), the sum

$$K(u, u') = K^{(1)}(u, u') + K^{(2)}(u, u') = \sum_{f \in \mathscr{F}_1 \cup \mathscr{F}_2} \frac{1}{\|f\|_{\mathscr{P}(\mathbb{R})}^2} f(u) f(u')$$

produces a kernel for direct sum $\mathscr{P}_{2e}(\mathbb{R}) = \mathscr{P}_{2e-1}^*(\mathbb{R}) \oplus \mathscr{P}_{2e}^*(\mathbb{R})$.

The following fact reveals a close connection between kernel functions and cubature formulas:

Theorem 1.3 (Hermite–Gauss quadrature, [6]) *Let H_e be the Hermite polynomial of degree e, which is orthogonal with respect to inner product $(\cdot, \cdot)_{\mathscr{P}(\mathbb{R})}$ as in Example 1.1, and let $x_1, \ldots, x_e$ be the simple roots of H_e. Further, let K be the kernel function corresponding to $\mathscr{P}_{2e-1}(\mathbb{R})$. Then*

$$\frac{1}{\int_{\mathbb{R}} \exp(-u^2) \, du} \int_{\mathbb{R}} f(u) \exp(-u^2) \, du = \sum_{\ell=1}^{e} \frac{1}{K(x_\ell, x_\ell)} f(x_\ell) \quad \text{for all } f \in \mathscr{P}_{2e-1}(\mathbb{R}).$$

The weights $1/K(x_\ell, x_\ell)$ are often called *Christoffel numbers*. The Christoffel numbers have various elegant expressions in terms of H_e and H_{e+1}. For example,

$$\frac{1}{K(x_\ell, x_\ell)} = -\frac{2^{e+1} e! \sqrt{\pi}}{H_{e+1}(x_\ell) H_e'(x_\ell)}, \tag{1.5}$$

where H_e' denotes the derivative of H_e. This directly follows from the so-called *Christoffel–Darboux formula* of K, as will become clear in the next section (see Example 1.4).

Mysovskikh [11] proves higher dimensional analogues of Theorem 1.3, which are fundamental in the theory of cubature formulas; for example, see Chap. 2. In Chap. 5, we will generalize the Mysovskikh result to hold for all finite-dimensional Hilbert spaces.

1.2   Kernels and Compact Formulas

Let $\mathscr{H}$ be a real vector space with inner product $(\cdot, \cdot)_{\mathscr{H}}$. A subset $\{f_i\}_i$ of $\mathscr{H}$ is called an *orthonormal system (ONS)* if $(f_i, f_j)_{\mathscr{H}} = \delta_{ij}$. An orthogonal system $\{f_i\}_i$ of $\mathscr{H}$ is called an *orthonormal basis (ONB)* if

$$(f, f_i)_{\mathscr{H}} = 0 \quad \text{for all } i \geq 1 \quad \Longrightarrow \quad f = 0.$$

For any Hilbert space, there exists an orthonormal basis consisting of at most countably many vectors. For a basis of $\mathscr{H}$, an orthonormal basis can be constructed by using the Gram–Schmidt orthonormalization. The following fact is standard.

Proposition 1.2 *Let $\{f_i\}_i$ be a subset of a real Hilbert space $\mathscr{H}$ with inner product $(\cdot, \cdot)_{\mathscr{H}}$. Then the following are equivalent:*

(i) $\{f_i\}_i$ *is an orthonormal basis.*
(ii) $\mathrm{Span}_{\mathbb{R}}\{f_i\}_i$ *is dense in* $\mathscr{H}$.
(iii) $f = \sum_i (f, f_i)_{\mathscr{H}} f_i$ *for all* $f \in \mathscr{H}$ *(Fourier expansion).*
(iv) $\|f\|_{\mathscr{H}}^2 = \sum_i |(f, f_i)_{\mathscr{H}}|^2$ *for all* $f \in \mathscr{H}$ *(Parseval identity).*

Let $\mathscr{H}$ be a real Hilbert space of $\mathbb{R}$-functions on a set Ω, with inner product $(\cdot, \cdot)_{\mathscr{H}}$. For each $\omega \in \Omega$, the linear functional L_ω is defined as

$$L_\omega : \mathscr{H} \longrightarrow \mathbb{R}, \quad f \longmapsto f(\omega). \tag{1.6}$$

Moreover, assume that for each ω, L_ω is bounded.[3] This is equivalent to say that for each $\omega \in \Omega$, L_ω is continuous. This condition is automatically satisfied when $\mathscr{H}$ has finite dimension.

Remark 1.3 For a special type of integrals, called centrally symmetric integrals (see Sect. 2.3), Möller [9] shows a lower bound for the number of points in a given cubature formula of odd degree and characterizes cubature formulas which are minimal with respect to his bound. Note that the functional L_0 plays a key role in his results; see Theorem 2.3 of Chap. 2. The Möller work has not been recognized in the theory of Euclidean designs in the design of experiments and algebraic combinatorics, and Bannai et al. [3] first bring the idea of Möller into the theory of Euclidean design and thereby improves a classical lower bound for Euclidean designs. More details will be explained in Sect. 2.2.

The following fact is standard in functional analysis.

Theorem 1.4 (Riesz representation theorem) *With the above* L_ω, *there exists a unique function* $f_\omega \in \mathscr{H}$ *such that*

$$L_\omega(f) = (f, f_\omega)_{\mathscr{H}} \text{ for every } f \in \mathscr{H}.$$

Moreover it holds that

$$\inf\{M \geq 0 \mid |L_\omega(f)| \leq M\|f\|_{\mathscr{H}} \text{ for all } f \in \mathscr{H}\} = \|f_\omega\|_{\mathscr{H}}.$$

The following fact mentions an approach for explicitly realizing kernel functions.

Proposition 1.3 *With* Ω, f_ω, $\mathscr{H}$ *as in Theorem 1.4, take an orthonormal basis* $\{f_i\}_i$ *of* $\mathscr{H}$. *Let* K *be the kernel function corresponding to* $\mathscr{H}$. *Then*

$$K(\omega, \omega') = \sum_j f_j(\omega) f_j(\omega') \text{ for all } \omega, \omega' \in \Omega.$$

---

[3]For real normed vector spaces, $\mathscr{H}$ and $\mathscr{H}'$, an $\mathbb{R}$-linear map $L : \mathscr{H} \to \mathscr{H}'$ is said to be *bounded* if there exists some $M > 0$ such that $\|L(f)\|_{\mathscr{H}'} \leq M\|f\|_{\mathscr{H}}$ for all $f \in \mathscr{H}$.

Proof Let $\omega, \omega' \in \Omega$. Then it follows from Proposition 1.2 (iii) and Theorem 1.4 that

$$
\begin{aligned}
K(\omega, \omega') &= (f_\omega, f_{\omega'})_{\mathscr{H}} \\
&= \left(\sum_j (f_\omega, f_j)_{\mathscr{H}} f_j, \sum_j (f_{\omega'}, f_j)_{\mathscr{H}} f_j \right)_{\mathscr{H}} \\
&= \sum_i (f_\omega, f_i)_{\mathscr{H}} (f_{\omega'}, f_i)_{\mathscr{H}} \\
&= \sum_i f_i(\omega) f_i(\omega')
\end{aligned}
$$

which completes the proof. □

It is realized, in (1.5), that Christoffel numbers of the Hermite–Gauss quadrature have an elegant expression in terms of Hermite polynomials. To understand it, the following fact, Example 1.4, is needed (e.g., [6]).

Example 1.4 (*Christoffel–Darboux formula*) Let $\mathscr{H} = \mathscr{P}_t(\mathbb{R})$, and consider Gaussian integration $\mathscr{I}[f] := \int_{\mathbb{R}} f(u) \exp(-u^2)\, du / \int_{\mathbb{R}} \exp(-u^2)\, du$. The Hermite polynomial of degree ℓ, say $H_\ell(x)$, is an orthogonal polynomial with respect to $\mathscr{I}$. By (1.4), the kernel K for polynomial space $\mathscr{H}$ is then given by

$$
K(u, v) = \sum_{\ell=0}^{e} \frac{1}{\|H_\ell\|_{\mathscr{H}}^2} H_\ell(u) H_\ell(v).
$$

This has the following "compact form", called *Christoffel–Darboux formulas*:

$$
K(u, u') = \frac{k_t}{h_t k_{t+1}} \cdot \frac{H_t(u') H_{t+1}(u) - H_t(u) H_{t+1}(u')}{u - u'}, \quad u, u' \in \mathbb{R},
$$

$$
K(u, u) = \frac{k_t}{h_t k_{t+1}} (H'_{t+1}(u) H_t(u) - H'_t(u) H_{t+1}(u)), \quad u \in \mathbb{R} \tag{1.7}
$$

where H_ℓ has the leading coefficient k_ℓ and squared norm h_ℓ.

To see (1.5), the formula (1.7) is used. Using the fact (e.g., [17]) that $k_t = 2^t$ and $h_t = 2^t t! \sqrt{\pi}$, it follows that

$$
\frac{1}{K(x_\ell, x_\ell)} = \frac{h_t k_{t+1}}{k_t H_{t+1}(x_\ell) H'_t(x_\ell)} = -\frac{2^{t+1} t! \sqrt{\pi}}{H'_t(x_\ell) H_{t+1}(x_\ell)}.
$$

In general, the kernel function admits the Christoffel–Darboux formula for any sequence of orthogonal polynomials defined on a subset of $\mathbb{R}$. However, in higher dimensions, the situation is quite different, with some special exceptions creating beautiful connections with the theory of cubature formulas. The details on these connections will be explained throughout this book.

Compact formulas are here described for kernels of several linear spaces of polynomials defined on $(d-1)$-dimensional unit sphere $\mathbb{S}^{d-1} = \{\omega \in \mathbb{R}^d \mid \|\omega\| = 1\}$ with $\|\omega\| = \sqrt{\langle \omega, \omega \rangle}$.

The *Gegenbauer polynomial of degree ℓ with parameter $\alpha > -1/2$*, say $C_\ell^{(\alpha)}$ is defined by the generating function

$$\frac{1}{(1 + 2ux + x^2)^\alpha} = \sum_{\ell=0}^{\infty} C_\ell^{(\alpha)}(u)x^\ell. \tag{1.8}$$

Polynomials $C_\ell^{(\alpha)}(u)$ are orthogonal with respect to measure $(1-u^2)^{\alpha-1/2}\, du$ defined on interval $[-1, 1]$; for example, see Erdélyi et al. [7, pp. 174–178]. For further arguments, the case where $\alpha = (d-2)/2$ is only considered.

Let $\mathrm{Hom}_\ell(\mathbb{R}^d)$ be the space of all homogeneous polynomials of degree exactly ℓ in d variables $\omega_1, \ldots, \omega_d$. Also denote by $\mathrm{Harm}_\ell(\mathbb{R}^d)$ the subspace of harmonic polynomials, namely

$$\mathrm{Harm}_\ell(\mathbb{R}^d) = \left\{ f \in \mathrm{Hom}_\ell(\mathbb{R}^d) \mid \sum_{i=1}^{d} \frac{\partial^2}{\partial \omega_i^2} f = 0 \right\}.$$

It is well known (cf. [7, 16]) that

$$h_\ell^d := \dim \mathrm{Harm}_\ell(\mathbb{R}^d) = \dim \mathrm{Harm}_\ell(\mathbb{S}^{d-1}) = \binom{d+\ell-1}{\ell} - \binom{d+\ell-3}{\ell-2}. \tag{1.9}$$

The notation $\mathrm{Harm}_\ell(\mathbb{S}^{d-1})$ is used for the space of all polynomials in $\mathrm{Harm}_\ell(\mathbb{R}^d)$ restricted to sphere $\mathbb{S}^{d-1}$, and similarly for other polynomial spaces and regions Ω.

Example 1.5 Here are small examples of $C_\ell^{((d-2)/2)}$:

$$C_0^{((d-2)/2)}(u) \equiv 1, \quad C_1^{((d-2)/2)}(u) = (d-2)u,$$

$$C_2^{((d-2)/2)}(u) = \frac{(d-2)(du^2-1)}{2}, \quad C_3^{((d-2)/2)}(u) = \frac{d(d-2)u\{(d+2)u^2-3\}}{3!},$$

$$C_4^{((d-2)/2)}(u) = \frac{(d-2)d\{(d+2)(d+4)u^4 - 6(d+2)u^2 + 3\}}{4!}.$$

By (1.9), it is checked that

$$h_0^d = 1, \quad h_1^d = d, \quad h_2^d = \frac{1}{2}(d-1)(d+2), \quad h_3^d = \frac{1}{6}(d-1)d(d+4).$$

The following fact, called the *addition formula* in spherical harmonics (e.g., Erdélyi et al. [7, pp. 242–243]), plays a crucial role in the study of cubature formulas for spherical integration.

Theorem 1.5 (Addition formula) *Let $\{\phi_{\ell,i} \mid 1 \leq i \leq h_{\ell}^d\}$ be an orthonormal basis of* $\mathrm{Harm}_{\ell}(\mathbb{S}^{d-1})$ *with respect to inner product*

$$(f, g)_{\mathbb{S}^{d-1}} = \frac{1}{|\mathbb{S}^{d-1}|} \int_{\mathbb{S}^{d-1}} f(\omega) g(\omega) \, \rho(d\omega)$$

where $|\mathbb{S}^{d-1}|$ is the surface area and ρ is the surface measure on $\mathbb{S}^{d-1}$. Then, with h_{ℓ}^d defined in (1.9),

$$\sum_{i=1}^{h_{\ell}^d} \phi_{\ell,i}(\omega) \phi_{\ell,i}(\omega') = \frac{d + 2\ell - 2}{d - 2} C_{\ell}^{((d-2)/2)}\left(\langle \omega, \omega' \rangle\right), \quad \omega, \omega' \in \mathbb{S}^{d-1}.$$

Remark 1.4 In combinatorics and related areas, the notation Q_{ℓ} or Q_{ℓ}^d is more widely used to mean the scaled Gegenbauer polynomial of degree ℓ, namely

$$Q_{\ell}(u) = Q_{\ell}^d(u) = \frac{d + 2\ell - 2}{d - 2} C_{\ell}^{((d-2)/2)}(u), \quad u \in \mathbb{R}.$$

The following is another famous example of compact formulas [18].

Theorem 1.6 *Denote by $\mathbb{B}^d$ the unit ball centered at the origin, namely, $\mathbb{B}^d := \{\omega \in \mathbb{R}^d \mid \|\omega\| \leq 1\}$. Let $K_t^{(\alpha)}$ be the kernel for polynomial space $\mathscr{P}_t(\mathbb{B}^d)$ with respect to multivariate Jacobi integral $\int_{\mathbb{B}^d} \cdot (1 - \|\omega\|^2)^{\alpha - 1/2} d\omega / \int_{\mathbb{B}^d} (1 - \|\omega\|^2)^{\alpha - 1/2} d\omega$, where $\alpha \geq 0$. Then*

$$
\begin{aligned}
&K_t^{(\alpha)}(\omega, \omega') \\
&= \int_0^{\pi} \left\{ C_t^{(\alpha + \frac{d+1}{2})}\left(\langle \omega, \omega' \rangle + \sqrt{1 - \|\omega\|^2}\sqrt{1 - \|\omega'\|^2} \cos \psi \right) \right. \\
&\quad \left. + C_{t-1}^{(\alpha + \frac{d+1}{2})}\left(\langle \omega, \omega' \rangle + \sqrt{1 - \|\omega\|^2}\sqrt{1 - \|\omega'\|^2} \cos \psi \right) \right\} \\
&\quad \times (\sin \psi)^{2\alpha - 1} d\psi \Big/ \int_0^{\pi} (\sin \psi)^{2\alpha - 1} \, d\psi \\
&= \frac{2\Gamma(\alpha + \frac{d+2}{2}) \Gamma(t + 2\alpha + d)}{\Gamma(2\alpha + d + 1) \Gamma(t + \alpha + d/2)} \\
&\quad \times \int_0^{\pi} P_t^{(\alpha + d/2, \alpha + d/2 - 1)}\left(\langle \omega, \omega' \rangle + \sqrt{1 - \|\omega\|^2}\sqrt{1 - \|\omega'\|^2} \cos \psi \right) \\
&\quad \times (\sin \psi)^{2\alpha - 1} d\psi \Big/ \int_0^{\pi} (\sin \psi)^{2\alpha - 1} \, d\psi, \quad \omega, \omega' \in \mathbb{B}^d
\end{aligned}
$$

and, for $\alpha = 0$,

$$K_t^{(0)}(\omega, \omega')$$

$$= \frac{1}{2}\left\{ C_t^{(\frac{d+1}{2})}\left(\langle\omega, \omega'\rangle + \sqrt{1 - \|\omega\|^2}\sqrt{1 - \|\omega'\|^2}\right) + C_{t-1}^{(\frac{d+1}{2})}\left(\langle\omega, \omega'\rangle + \sqrt{1 - \|\omega\|^2}\sqrt{1 - \|\omega'\|^2}\right)\right\}$$

$$+ \frac{1}{2}\left\{ C_m^{(\frac{d+1}{2})}\left(\langle\omega, \omega'\rangle - \sqrt{1 - \|\omega\|^2}\sqrt{1 - \|\omega'\|^2}\right) + C_{t-1}^{(\frac{d+1}{2})}\left(\langle\omega, \omega'\rangle - \sqrt{1 - \|\omega\|^2}\sqrt{1 - \|\omega'\|^2}\right)\right\}$$

$$= \frac{\Gamma(\frac{d+2}{2})\Gamma(t+d)}{\Gamma(d+1)\Gamma(t+d/2)}\left\{ P_t^{(d/2, d/2-1)}\left(\langle\omega, \omega'\rangle + \sqrt{1 - \|\omega\|^2}\sqrt{1 - \|\omega'\|^2}\right)\right.$$

$$\left. + P_t^{(d/2, d/2-1)}\left(\langle\omega, \omega'\rangle - \sqrt{1 - \|\omega\|^2}\sqrt{1 - \|\omega'\|^2}\right)\right\}, \quad \omega, \omega' \in \mathbb{B}^d$$

where Γ is the Gamma function and $P_t^{(\beta, \beta')}$ is the Jacobi polynomial of degree t with parameters β, $\beta' > -1$.

Xu [19] uses the above formulas to derive a lower bound for the number of points in a given cubature on the unit ball. His bound belongs to a class of *linear programming (LP) bounds*, which, as will be discussed in Sect. 2.4 (see also Sect. 1.3), can be the best known bound in some special situations, where compact formulas given in Theorem 1.6 play a crucial role; for details see the original paper by Xu [19].

1.3 Kernels and Dimension

Let μ be a positive measure on a set Ω. Further let $\mathscr{L}^2(\Omega, \mu)$ be the space of all square-integrable $\mathbb{R}$-functions for $\int_\Omega \cdot d\mu / \int_\Omega d\mu$, namely, $\mathscr{L}^2(\Omega, \mu)$ consists of all $\mathbb{R}$-functions f such that

$$\left(\frac{1}{\int_\Omega \mu(d\omega)}\int_\Omega |f(\omega)|^2\, \mu(d\omega)\right)^{1/2} < \infty. \tag{1.10}$$

Consider a subset of $L^2(\Omega, \mu)$ defined by

$$\mathscr{Z} = \{f \in \mathscr{L}^2(\Omega, \mu) \mid f = 0 \text{ at almost everywhere with respect to } \mu\}.$$

Clearly, $\mathscr{Z}$ is a subspace of $\mathscr{L}^2(\Omega, \mu)$.

Let $\mathscr{H}$ be a subspace of $\mathscr{L}^2(\Omega, \mu)$ which is embedded in the quotient space $L^2(\Omega, \mu) = \mathscr{L}^2(\Omega, \mu)/\mathscr{Z}$. Let $(\cdot, \cdot)_{\mathscr{H}}$ be the inner product corresponding to (1.10). When $\mathscr{F} \subset \mathscr{H}$ is an orthonormal basis of $\mathscr{H}$, the cardinality of $\mathscr{F}$ is called the *dimension* of $\mathscr{H}$ and denoted by $\dim \mathscr{H}$.

Proposition 1.4 *With the above notation Ω, μ, $\mathscr{H}$, let K be the kernel corresponding to $\mathscr{H}$. Moreover, assume that μ is finite, $\mathscr{H}$ is closed with respect to $(\cdot, \cdot)_{\mathscr{H}}$, and $\int_{\Omega \times \Omega} |K(\omega, \omega')|^2\, \mu(d\omega)\mu(d\omega') < \infty$. Then $\mathscr{H}$ has finite dimension and*

$$\dim \mathscr{H} = \frac{1}{\int_\Omega \mu(d\omega)} \int_\Omega K(\omega, \omega)\, \mu(d\omega).$$

Proof Since $\int_{\Omega \times \Omega} |K(\omega, \omega')|^2\, \mu(d\omega)\mu(d\omega') < \infty$, K is a Hilbert–Schmidt kernel that corresponds to the Hilbert–Schmidt integral operator

$$F_K : L^2(\Omega, \mu) \longrightarrow L^2(\Omega, \mu), \quad f(\cdot) \longmapsto \frac{1}{\int_\Omega \mu(d\omega)} \int_\Omega K(\cdot, \omega') f(\cdot)\, \mu(d\omega');$$

see, e.g., [1]. Since K has the reproducing property, F_K is the identity map which is not a Hilbert–Schmidt operator in the infinite case. Therefore, the dimension of $\mathscr{H}$ is finite. Finally, let $\{f_i\}_{i=1}^{\dim \mathscr{H}}$ be an orthonormal basis of $\mathscr{H}$. Then it follows from Proposition 1.3 that

$$\dim \mathscr{H} = \sum_{i=1}^{\dim \mathscr{H}} \frac{1}{\int_\Omega \mu(d\omega)} \int_\Omega f_i^2(\omega)\, \mu(d\omega) = \frac{1}{\int_\Omega \mu(d\omega)} \int_\Omega \sum_i f_i^2(\omega)\, \mu(d\omega)$$

$$= \frac{1}{\int_\Omega \mu(d\omega)} \int_\Omega K(\omega, \omega)\, \mu(d\omega). \qquad \square$$

Example 1.6 Let K be the kernel function that corresponds to the space, Harm_ℓ $(\mathbb{S}^{d-1})$, of all harmonic homogeneous polynomials of degree exactly ℓ on unit sphere $\mathbb{S}^{d-1}$. Then, by the addition formula (Theorem 1.5) and Remark 1.4, the dimension of $\mathrm{Harm}_\ell(\mathbb{S}^{d-1})$ is expressed in terms of Gegenbauer polynomial Q_ℓ as follows:

$$K(\omega, \omega') = Q_\ell^{((d-2)/2)}\left(\langle \omega, \omega' \rangle\right), \quad \omega, \omega' \in \mathbb{S}^{d-1}.$$

For each $\omega \in \mathbb{S}^{d-1}$, consider the function $f_\omega : \mathbb{S}^{d-1} \to \mathbb{R}$ given by

$$f_\omega(\omega') = Q_\ell(\langle \omega', \omega \rangle).$$

By the Riesz representation theorem (Theorem 1.4), it is seen that

$$K(\omega, \omega') = (f_\omega, f_{\omega'})_{\mathbb{S}^{d-1}} = f_{\omega'}(\omega) = Q_\ell(\langle \omega, \omega' \rangle). \tag{1.11}$$

Here recall the notation $(f, g)_{\mathbb{S}^{d-1}} := \int_{\mathbb{S}^{d-1}} fg\, d\rho / |\mathbb{S}^{d-1}|$.

By use of Theorem 1.7, it is concluded that $h_\ell^d = Q_\ell(1)$, since

$$\dim \mathrm{Harm}_\ell(\mathbb{S}^{d-1}) = \frac{1}{|\mathbb{S}^{d-1}|} \int_{\mathbb{S}^{d-1}} K(\omega, \omega)\, \rho(d\omega)$$

$$= \frac{1}{|\mathbb{S}^{d-1}|} \int_{\mathbb{S}^{d-1}} Q_\ell(\langle \omega, \omega \rangle)\, \rho(d\omega)$$

$$= \frac{1}{|\mathbb{S}^{d-1}|} \int_{\mathbb{S}^{d-1}} Q_\ell(1)\, \rho(d\omega)$$

$$= Q_\ell(1).$$

In general, there are various expressions to find the values of classical orthogonal polynomials $\Phi(u)$ at $u = 1$. For instance, Szegő [17, p.80] covers the desired information for Jacobi polynomials. Among various expressions for $Q_\ell(1)$, it seems that

$$Q_\ell(1) = \binom{d+\ell-1}{\ell} - \binom{d+\ell-3}{\ell-2} \tag{1.12}$$

is preferred in combinatorics and harmonic analysis.

Example 1.7 (*Example* 1.6, *continued*) The famous Fischer decomposition states (e.g., [16]) that

$$\mathrm{Hom}_\ell(\mathbb{S}^{d-1}) = \bigoplus_{\substack{0 \le i \le \ell \\ i \equiv \ell \ (\mathrm{mod}\ 2)}} \mathrm{Harm}_i(\mathbb{S}^{d-1}),$$

$$\mathscr{P}_\ell(\mathbb{S}^{d-1}) = \bigoplus_{0 \le i \le \ell} \mathrm{Harm}_i(\mathbb{S}^{d-1}). \tag{1.13}$$

By combining (1.11) with Theorem 1.2, kernel functions for spaces $\mathrm{Hom}_\ell(\mathbb{S}^{d-1})$ and $\mathscr{P}_\ell(\mathbb{S}^{d-1})$, say K_{Hom} and $K_{\mathscr{P}}$ respectively, are given by

$$K_{\mathrm{Hom}}(\omega, \omega') = \sum_{\substack{0 \le i \le \ell \\ i \equiv \ell \ (\mathrm{mod}\ 2)}} Q_i(\langle \omega, \omega' \rangle),$$

$$K_{\mathscr{P}}(\omega, \omega') = \sum_{0 \le i \le \ell} Q_i(\langle \omega, \omega' \rangle). \tag{1.14}$$

Some standard arguments in calculus (e.g., [17]) show that polynomials

$$P_\ell(u) := \sum_{\substack{0 \le i \le \ell, \\ i \equiv \ell \ (\mathrm{mod}\ 2)}} Q_i(u), \quad R_\ell(u) := \sum_{0 \le i \le \ell} Q_i(u), \quad \ell \ge 0 \tag{1.15}$$

are orthogonal with respect to measures $(1-u)^{d/2}(1+u)^{(d-2)/2} du$ and $(1-u^2)^d du$ on interval $[-1, 1]$, respectively. It thus follows from Theorem 1.7 and (1.12) that

$$\dim \mathrm{Hom}_\ell(\mathbb{S}^{d-1}) = P_\ell(1)$$

$$= \sum_{\substack{0 \le i \le \ell \\ i \equiv \ell \pmod 2}} Q_i(1)$$

$$= \sum_{\substack{0 \le i \le \ell \\ i \equiv \ell \pmod 2}} \left\{ \binom{d+i-1}{i} - \binom{d+i-3}{i-2} \right\}$$

$$= \binom{d+\ell-1}{\ell}. \tag{1.16}$$

Similar arguments also show that

$$\dim \mathscr{P}_\ell(\mathbb{S}^{d-1}) = \binom{d+\ell-1}{\ell} + \binom{d+\ell-2}{\ell-1}, \tag{1.17}$$

which is used to derive a certain class of bounds for cubature formulas including the following result.

Theorem 1.7 *Assume that there exist points* $x_1, \ldots, x_n \in \mathbb{S}^{d-1}$ *and positive real numbers* $w_1, \ldots, w_n$ *such that*

$$\frac{1}{|\mathbb{S}^{d-1}|} \int_{\mathbb{S}^{d-1}} f(\omega)\, \rho(d\omega) = \sum_{i=1}^n w_i f(x_i) \quad \text{for all } f \in \mathscr{P}_t(\mathbb{S}^{d-1}).$$

Then the number of points is bounded below as

$$n \ge \begin{cases} \dbinom{d+e-1}{e} + \dbinom{d+e-2}{e-1} & \text{if } t = 2e, \\[3mm] 2\dbinom{d+e-1}{e} & \text{if } t = 2e+1. \end{cases}$$

This bound is a kind of *Fisher-type bounds*, which is substantially different from LP bounds that have been briefly stated at the last part of Sect. 1.2.

1.4   Further Remarks

The quadratic forms (the left-hand side of (1.1)) appearing in the definition of kernel functions in Sect. 1.1 are of interest in its own, providing interdisciplinary topics in fields as diverse as combinatorics, geometry, coding theory, electrostatic, etc., where the term "potential energy" is frequently used for those quadratic forms. For example, the following fact, Theorem 1.8 of [14], is well known.

Theorem 1.8 (Seidel'nikov inequality) *Let X be a finite subset of unit sphere $\mathbb{S}^{d-1}$. Then for each positive integer ℓ*

$$\frac{1}{|X|^2} \sum_{x,y \in X} \langle x, y \rangle^\ell$$

$$\geq \begin{cases} \dfrac{1}{|\mathbb{S}^{d-1}|^2} \displaystyle\int_{\mathbb{S}^{d-1} \times \mathbb{S}^{d-1}} \langle \omega, \omega' \rangle^\ell \, \rho(d\omega)\rho(d\omega') & \text{if } \ell \equiv 0 \pmod 2, \\ 0 & \text{if } \ell \equiv 1 \pmod 2. \end{cases}$$

Moreover, the equality holds for all $1 \leq \ell \leq t$ if and only if X is a spherical t-design on $\mathbb{S}^{d-1}$.

Roughly speaking, a *spherical design* in Theorem 1.8 coincides with an equi-weighted cubature formula for spherical integration $\int_{\mathbb{S}^{d-1}} \cdot \, d\rho / |\mathbb{S}^{d-1}|$. The precise definition will be given in Chap. 2, together with that of *Euclidean designs*.

It is interesting to discuss Seidel'nikov-type inequalities for other kernel functions. For example, consider $K(\omega, \omega') := Q_\ell(\langle \omega, \omega' \rangle)$ as in Example 1.6. Let $x_1, \ldots, x_n \in \mathbb{S}^{d-1}$ with positive weights $w_1, \ldots, w_n$. Then it follows that

$$\sum_{i,j=1}^{n} w_i w_j Q_\ell(\langle x_i, x_j \rangle) \geq 0. \tag{1.18}$$

Recall that $Q_\ell(\langle \omega, \omega' \rangle)$ is a kernel function corresponding to $\mathrm{Harm}_\ell(\mathbb{S}^{d-1})$. To find conditions for equality, a key observation is that

$$\sum_{i,j=1}^{n} w_i w_j Q_\ell(\langle x_i, x_j \rangle) = \sum_{i,j=1}^{n} w_i w_j \sum_{k=1}^{h_\ell^d} \phi_k(x_i)\phi_k(x_j)$$

$$= \sum_{k=1}^{h_\ell^d} \left(\sum_{i=1}^{n} w_i \phi_k(x_i) \right)^2$$

where $\{\phi_k\}_{k=1}^{h_\ell^d}$ is an orthonormal basis of $\mathrm{Harm}_i(\mathbb{S}^{d-1})$. Therefore, the equality holds in (1.18) if and only if

$$\sum_{k=1}^{n} w_k f(x_k) = 0 \quad \text{for all } f \in \mathrm{Harm}_\ell(\mathbb{S}^{d-1}).$$

Hence, the Fischer decomposition (1.13) yields the following result.

Theorem 1.9 *Let $x_1, \ldots, x_n \in \mathbb{S}^{d-1}$ with positive weights $w_1, \ldots, w_n$. Then the following are equivalent:*

(i) $\sum_{i,j=1}^{n} w_i w_j Q_\ell(\langle x_i, x_j \rangle) = 0$ *for all $1 \leq \ell \leq t$.*

(ii) x_i *and* w_i, $1 \leq i \leq n$, *form a cubature formula of degree t for spherical integration, namely,* $\int_{\mathbb{S}^{d-1}} f(\omega)\, \rho(d\omega)/|\mathbb{S}^{d-1}| = \sum_{i=1}^{n} w_i f(x_i)$ *for all* $f \in \mathscr{P}_t(\mathbb{S}^{d-1})$.

Now, Proposition 1.3 in Sect. 1.2 gives a construction of kernel functions K by utilizing orthonormal basis of the corresponding Hilbert spaces. We shall look at a different construction, which is related to kernel methods in machine learning; for example, see [5, 10].

A kernel function $K : \mathbb{R}^d \times \mathbb{R}^d \to \mathbb{R}$ is said to be *shift invariant*, if there exists a univariate function ϕ such that

$$K(\omega, \omega') = \phi(\omega - \omega') \quad \text{for all } \omega, \omega' \in \mathbb{R}^d.$$

For example, the Gaussian kernel $\exp(-\|\omega - \omega'\|^2/2\sigma^2)$ is in a class of shift-invariant kernels.

Theorem 1.10 (Bochner theorem (e.g., [12])) *Let ϕ be a continuous function on $\mathbb{R}^d$ such that $\phi(0) = 1$. Then the following are equivalent:*

(i) ϕ is positive semi-definite, namely

$$\sum_{i,j=1}^{n} c_i c_j \phi(\omega_i - \omega_j) \geq 0 \quad \textit{for all } c_1, \ldots, c_n \in \mathbb{R} \textit{ and all } \omega_1, \ldots, \omega_n \in \mathbb{R}^d.$$

(ii) There exists a nonnegative probability measure μ on $\mathbb{R}^d$ such that

$$\phi(\omega) = \int_{\mathbb{R}^d} \exp(\sqrt{-1}\tilde{\omega}^T \omega)\mu(d\tilde{\omega}).$$

In particular, any continuous shift-invariant kernel K is the Fourier transform of some probability measure μ, namely

$$K(\omega, \omega') = \int_{\mathbb{R}^d} \exp(\sqrt{-1}z^T(\omega - \omega'))\, \mu(dz). \tag{1.19}$$

The *random Fourier features*, which approximate integral (1.19) by using randomly chosen n samples, present a common approach for finding a feature map from $\mathbb{R}^d$ to $\mathbb{C}^n$. Dao et al. [5] propose a different deterministic approach by considering the feature map

$$f(\omega) = (\sqrt{w_1}\exp(\sqrt{-1}\omega_1^T \omega), \ldots, \sqrt{w_n}\exp(\sqrt{-1}\omega_n^T \omega)),$$

where $\omega_1, \ldots, \omega_n$ are fixed points in $\mathbb{R}^d$ with fixed weights $w_1, \ldots, w_n > 0$. Our objective is then to evaluate the approximation error

$$\varepsilon := \sup_{\|\omega\| \leq M} \left| \int_{\mathbb{R}^d} \exp(\sqrt{-1}z^T \omega)\mu(dz) - \sum_{i=1}^{n} w_i \exp(\sqrt{-1}\omega_i^T \omega) \right|$$

where M is the supremum of the usual Euclidean norm $\|\omega - \omega'\|$ among all possible points ω, ω' in a given region $\mathscr{M}$. Dao et al. [5] investigate the error for many classes of cubature formulas such as *Smolyak method* [15] and *product rule* (which they also call "polynomially-exact rules"); for more details, see Sect. 2.1. They also show good performance of their methods compared to the random Fourier features in some situations. For the details, we refer the reader to [5] and references therein.

References

1. Akhiezer, N.I., Glazman, I.M.: Theory of Linear Operators in Hilbert Space. Dover Publications, Inc., New York (1993). Translated from the Russian and with a preface by Merlynd Nestell. Reprint of the 1961 and 1963 translations. Two volumes bound as one
2. Aronszajn, N.: Theory of reproducing kernels. Trans. Am. Math. Soc. **68**, 337–404 (1950)
3. Bannai, E., Bannai, E., Hirao, M., Sawa, M.: Cubature formulas in numerical analysis and Euclidean tight designs. Eur. J. Comb. **31**(2), 423–441 (2010)
4. Bishop, C.M.: Pattern Recognition and Machine Learning. Springer, New York (2006)
5. Dao, T., De Sa, C., Ré, C.: Gaussian quadrature for kernel features. Adv. Neural. Inf. Process. Syst. **30**, 6109–6119 (2017)
6. Dunkl, C.F., Xu, Y.: Orthogonal Polynomials of Several Variables, Encyclopedia of Mathematics and its Applications, vol. 155, second edn. Cambridge University Press (2014)
7. Erdélyi, A., Magnus, W., Oberhettinger, F., Tricomi, F.: Higher Transcendental Functions, II. MacGraw-Hill (1953)
8. Kiefer, J., Wolfowitz, J.: The equivalence of two extremum problems. Can. J. Math. **12**, 363–366 (1960)
9. Möller, H.M.: Lower bounds for the number of nodes in cubature formulae. In: Numerische integration (Tagung, Math. Forschungsinst., Oberwolfach, 1978), Internat. Ser. Numer. Math., vol. 45, pp. 221–230. Birkhäuser, Basel-Boston, Mass. (1979)
10. Munkhoeva, M., Kapushev, Y., Oseledets, E.B.I.: Quadrature-based features for kernel approximation. Adv. Neural. Inf. Process. Syst. **31**, 9147–9156 (2018)
11. Mysovskih, I.P.: On the construction of cubature formulas with the smallest number of nodes (in Russian). Dokl. Akad. Nauk SSSR **178**, 1252–1254 (1968)
12. Rudin, W.: Fourier Analysis on Groups. Wiley (1990)
13. Saitoh, S.: Theory of Reproducing Kernels and its Applications. Pitman Research Notes in Mathemtical Series, 189. Longman Scientific and Technical, UK (1988)
14. Seidel'nikov, V.M.: New estimates for the closest packing of spheres in n-dimensional Euclidean space (in Russian). Mat. Sb. (N.S.) **95**, 148–158 (1974)
15. Smolyak, S.A.: Quadrature and interpolation formulas for tensor products of certain classes of functions (in Russian). Dokl. Akad. Nauk SSSR **148**(5), 1042–1053 (1963)
16. Stein, E.M., Weiss, G.: Introduction to Fourier Analysis on Euclidean Spaces. Princeton University Press (1971)
17. Szegő, G.: Orthogonal Polynomials. Colloquium Publications, Vol. XXIII. American Mathematical Society, Providence, R.I. (1975)
18. Xu, Y.: Summability of fourier orthogonal series for Jacobi weight on a ball in R^d. Trans. Am. Math. Soc. **351**(6), 2439–2458 (1999)
19. Xu, Y.: Lower bound for the number of nodes of cubature formulae on the unit ball. J. Complexity **19**(3), 392–402 (2003)

Chapter 2
Cubature Formula

A *cubature formula* reveals a numerical integration rule that approximates a multiple integral by a positive linear combination of function values at finitely many specified points on the integral domain. A central objective is to investigate the existence as well as the construction of cubature formulas in high dimensions. A great deal of work has been done on this subject from a viewpoint of numerical analysis, and several celebrated books, which the reader will find valuable introductions and references, are available; for example, see [13, 20, 21, 29, 57, 58, 61].

On the other hand, combinatorial and geometric configurations often appear in many examples of cubature formulas for the so-called *spherically symmetric integrals*, which are thus of great importance from a viewpoint of algebraic combinatorics and discrete geometry. In those areas of mathematics, the term *Euclidean design* is often used; e.g., see [2–4, 40]. The existence problem of Euclidean designs are closely related to topics as diverse as linear programming problems, quasi-Monte Carlo integrations, isometric embeddings of Banach spaces, and Hilbert identities.

The present chapter reviews basic results on the existence and construction of cubature formula, with particular emphasis on the relationship among cubature formulas and reproducing kernels introduced in Chap. 1, some of which involve further arguments in subsequent Chap. 3 through Chap. 5.

Section 2.1 first presents the precise definition of cubature formula and then discusses elementary construction methods. Section 2.2 outlines some basic results on the existence of cubature formulas including, a general existence theorem by Tchakaloff (Theorem 2.1), the *Fisher-type bound* which is a classical lower bound for the number of points in a cubature formula (Theorem 2.2), and the *Möller bound* (Theorem 2.3) which is an improvement of the Fisher-type bound. Section 2.3 gives the definition of Euclidean designs and some related results, for example, a remarkable theorem by Bondarenko et al. (Theorem 2.7), with which people in combinatorics and geometry have their familiarity. Finally, Sect. 2.4 is closed with advanced topics as such stated in the previous paragraph.

The original version of this chapter was revised: Missed out author corrections have been incorporated. The correction to this chapter is available at https://doi.org/10.1007/978-981-13-8075-4_6.

© The Author(s), under exclusive license to Springer Nature Singapore Pte Ltd. 2019
M. Sawa et al., *Euclidean Design Theory*, JSS Research Series in Statistics,
https://doi.org/10.1007/978-981-13-8075-4_2

2.1 Cubature Formula and Elementary Construction Methods

Let $\mathbb{R}^d$ be the d-dimensional Euclidean space with ordinary inner product $\langle \cdot, \cdot \rangle$ and norm $\| \cdot \|$. Throughout this chapter, let μ be a finite positive measure on a subset Ω of $\mathbb{R}^d$ unless otherwise noted. Given a function f on Ω, define the integral

$$\mathscr{I}[f] = \frac{1}{\int_\Omega \mu(d\omega)} \int_\Omega f(\omega)\mu(d\omega).$$

One of the central objectives in numerical analysis is to find a numerical integration formula that approximates $\mathscr{I}[f]$ by a weighted summation rule as

$$\mathscr{Q}[f] = \sum_{x \in X} w(x) f(x)$$

where X is a fixed finite subset of Ω and w is a positive weight function on X.

The definition of cubature formula is as follows.

Definition 2.1 (*Cubature formula*) Given a nonnegative integer t, a *cubature formula of degree t for $\mathscr{I}$* is defined by a finite weighted pair (X, w) such that

$$\mathscr{I}[f] = \mathscr{Q}[f] \quad \text{for all } f \in \mathscr{P}_t(\mathbb{R}^d).$$

In particular, the term *quadrature formula* is often used for 1-dimensional cubature formula.

The *Simpson rule*, for which there are many variations, is often used for numerical integrations. The following is the simplest one.

Example 2.1 (*Simpson rule*) Let $\mu(du) = du$ be the Lebesgue measure on closed interval $[a, b]$. Then

$$X = \left\{ a, \frac{b-a}{2}, b \right\}, \quad w(a) = w(b) = \frac{1}{6}, \quad w\left(\frac{b-a}{2}\right) = \frac{2}{3}$$

form a quadrature formula of degree 3 for $\int_a^b \cdot \, du/(b-a)$:

$$\frac{1}{b-a} \int_a^b f(u)\, du = \frac{1}{6} f(a) + \frac{2}{3} f\left(\frac{a+b}{2}\right) + \frac{1}{6} f(b) \quad \text{for all } f \in \mathscr{P}_3(\mathbb{R}).$$

Actually, it can be checked that

$$\frac{1}{b-a} \int_a^b u^\ell \, du = \frac{b^{\ell+1} - a^{\ell+1}}{(\ell+1)(b-a)} = \frac{a^\ell}{6} + \frac{2}{3}\left(\frac{a+b}{2}\right)^\ell + \frac{b^\ell}{6}, \quad \ell = 0, 1, 2, 3.$$

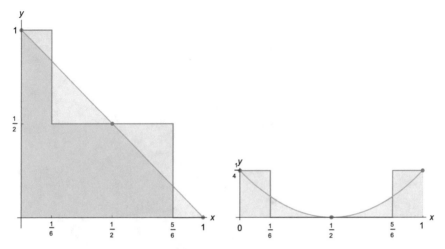

Fig. 2.1 Simpson rule for $f(u) = -u + 1$ (left) and $g(u) = (u - 1/2)^2$ (right)

Figure 2.1 illustrates the Simpson rule for definite integral $\int_0^1 \cdot \, du$.

Theorem 1.3 deals with a typical example of quadrature called *Hermite–Gauss quadrature*. For other Gaussian quadratures such as Jacobi–Gauss quadrature and Laguerre–Gauss quadrature, see e.g., Dunkl and Xu [20], Krylov [29] and Szegő [59]. Note that quadratures are utilized to characterize *optimal experimental designs* in Sects. 3.2 and 3.4.

Then what about the higher dimensional cases? It is not hard to construct cubature formulas for product measures on $\mathbb{R}^d$, e.g., for Gaussian measure $\mu(d\omega) = \exp(-\sum_{i=1}^d \omega_i^2) \, d\omega_1 \cdots d\omega_d = \prod_{i=1}^d \exp(-\omega_i^2) \, d\omega_i$, as the following example shows.

Example 2.2 (Product rule for Gaussian integration) Hermite–Gauss quadratures introduced in Theorem 1.3 are focused on again. For all integers a_j with $0 \le a_j \le 2e - 1$ and $j = 1, \ldots, d$,

$$
\frac{1}{\pi^{d/2}} \int_{\mathbb{R}^d} \omega_1^{a_1} \cdots \omega_d^{a_d} \exp\left(-\sum_{j=1}^d \omega_j^2 \right) d\omega_1 \cdots d\omega_d
$$

$$
= \prod_{j=1}^d \left(\frac{1}{\sqrt{\pi}} \int_{\mathbb{R}} \omega_j^{a_j} \exp\left(-\omega_j^2 \right) d\omega_j \right)
$$

$$
= \prod_{j=1}^d \left(\sum_{\ell_j=1}^e w_{\ell_j} x_{\ell_j}^{a_\ell} \right)
$$

$$
= \sum_{\ell_1=1}^e \cdots \sum_{\ell_d=1}^e \left(\prod_{j=1}^d w_{\ell_j} \right) x_{\ell_1}^{a_1} \cdots x_{\ell_d}^{a_d}
$$

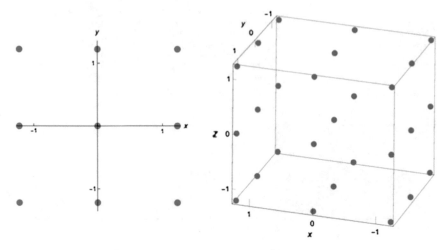

Fig. 2.2 Degree-five 3^2-points product rule for bivariate Gaussian integration (left), Degree-five 3^3-points product rule for trivariate Gaussian integration (right)

where K is the kernel polynomial corresponding to $\mathscr{P}_e(\mathbb{R})$ and $w_{\ell_j} = 1/K(x_{\ell_j}, x_{\ell_j})$. This is a cubature formula of degree $2e - 1$ for multivariate Gaussian integration $\mathscr{I}[\cdot] = \int_{\mathbb{R}^d} \cdot \exp(-\sum_{\ell=1}^d \omega_\ell^2) \, d\omega_1 \ldots d\omega_d/\pi^{d/2}$. Namely, it holds that for all $f \in \mathscr{P}_{2e-1}(\mathbb{R}^d)$,

$$\frac{1}{\pi^{d/2}} \int_{\mathbb{R}^d} f(\omega_1, \ldots, \omega_d) \exp\left(-\sum_{j=1}^d \omega_j^2\right) \, d\omega_1 \ldots d\omega_d$$

$$= \sum_{\ell_1=1}^e \cdots \sum_{\ell_d=1}^e \left(\prod_{j=1}^d w_{\ell_j}\right) f(x_{\ell_1}, \ldots, x_{\ell_d}).$$

See Fig. 2.2, where two point sets of product rules of degree 5 for Gaussian integration are illustrated in two and three dimensions, respectively.

Another typical method of constructing high-dimensional cubature formulas employs a particular type of group orbits, which works even for non-product measures.

Example 2.3 (cf. [62]) Let $\mathbb{S}^{d-1} = \{\omega \in \mathbb{R}^d \mid \|\omega\| = 1\}$ be the $(d-1)$-dimensional unit sphere with $d = 3k - 2$. Further let X be a set of all permutations and all sign changes of vector $(1/\sqrt{k}, 1/\sqrt{k}, \cdots, 1/\sqrt{k}, 0, \cdots, 0)$ such that the first k coordinates are equal to $1/\sqrt{k}$ and the remaining $2k - 2$ coordinates are equal to 0. Then it holds that

$$\frac{1}{|\mathbb{S}^{d-1}|} \int_{\mathbb{S}^{d-1}} f(\omega) \, \rho(d\omega) = \frac{1}{2^k \binom{3k-2}{k}} \sum_{x \in X} f(x) \quad \text{for all } f \in \mathscr{P}_5(\mathbb{R}^d),$$

where $|\mathbb{S}^{d-1}|$ is the surface area and ρ is the surface measure on $\mathbb{S}^{d-1}$.

Also, there are some other constructions based on reproducing kernels (see Theorems 2.4 and 2.5).

Two constructions mentioned above are quite simple and often appear in the context of numerical analysis and other related fields (cf. [16, 31]), which, however, suffer from serious drawback such that the resulting formulas have a larger number of points, as dimension d grows. This leads us to the following fundamental question.

Problem 2.1 *For $d \geq 2$, construct cubature formulas on $\mathbb{R}^d$ with a small number of points.*

Before going to Problem 2.1, it will be important to make clear what is meant by "small".

Problem 2.2 *Given $\mathscr{I}$, how small could the cardinality of X (say, $|X|$) in a cubature formula be?*

Partial solutions of Problem 2.2 are presented in the next section; see Theorems 2.2 and 2.3.

After lower bounds are obtained, the next goal is to find cubature formulas that are not far from these lower bounds. There are various methods, e.g., the *Victoir method* [62], the *Smolyak method* [56], the *Kuperberg method* [30], and so on, to overcome the "curse of dimensionality". Section 4.4 discusses such "thinning methods" as a generalization of the Victoir method.

2.2 Existence Theorems and Lower Bounds

A large amount of studies are available for the existence problem of cubature formula. Among them, a celebrated theorem by Tchakaloff [60], which is one of the fundamental results of the cubature theory, states that there always exist cubature formulas with sufficiently large cardinalities $|X|$. Several generalizations of Theorem 2.1 are available; see, e.g., Bayer and Teichmann [8], Curto and Fialkow [15] and Putiner [46].

Theorem 2.1 (Tchakaloff theorem) *For positive integers d and t, let μ be a positive measure on Ω in $\mathbb{R}^d$ satisfying $\int_\Omega |f(\omega)| \, \mu(d\omega) < \infty$ for all $f \in \mathscr{P}_t(\Omega)$. Then there exists a cubature formula of degree t for $\mathscr{I}$ with $X \subset \mathrm{supp}(\mu)$[1] and*

$$|X| \leq \dim \mathscr{P}_t(\Omega).$$

The *Carathéodory theorem* (e.g., [48, 55]) from convex analysis plays a key role to show Theorem 2.1. The Tchakaloff theorem will be extended to general functional spaces in Sect. 5.2.

---

[1] Let $\mathscr{B}$ be the Borel σ-algebra of Ω. Denote the support of μ by $\mathrm{supp}(\mu)$, i.e., closure of $\{B \in \mathscr{B} \mid \mu(B) > 0\}$.

Next let us look at two types of lower bounds for the number of points in a cubature formula. The lower bound in Theorem 2.2 is well known as the *Fisher-type bound* in combinatorics and related areas or the *Stroud bound* in analysis; see, e.g., Cools et al. [14], Delsarte and Seidel [18] and Stroud [58].

Theorem 2.2 (Fisher-type bound) *Let X be the set of points of a cubature formula of degree t for $\mathscr{I}$. Then the following holds:*

$$|X| \geq \dim \mathscr{P}_{\lfloor t/2 \rfloor}(\Omega). \tag{2.1}$$

A sketch of proof of this theorem (along the lines given in Stroud [58]) goes as follows.

Sketch of proof. A cubature formula of degree 4 for a bivariate Gaussian integral is focused for simplicity. A key point of the proof is to use the non-singularity of the moment matrix.

Assume that there exists an n-point cubature formula of the form

$$\frac{1}{\pi} \int_{\mathbb{R}^2} f(\omega_1, \omega_2) e^{-(\omega_1^2 + \omega_2^2)} d\omega_1 d\omega_2 = \sum_{\ell=1}^{n} w_\ell f(x_\ell, y_\ell) \quad \text{for all } f \in \mathscr{P}_4(\mathbb{R}^2).$$

Then by letting $f(\omega_1, \omega_2) = (1, \omega_1, \omega_2, \omega_1^2, \omega_1\omega_2, \omega_2^2)^T$, it holds that

$$(f(x_1, y_1)^T, \ldots, f(x_n, y_n)^T)^T \operatorname{diag}(w_1, \ldots, w_n)(f(x_1, y_1), \ldots, f(x_n, y_n))$$

$$= \begin{pmatrix} 1 & 0 & 0 & 1/2 & 0 & 1/2 \\ 0 & 1/2 & 0 & 0 & 0 & 0 \\ 0 & 0 & 1/2 & 0 & 0 & 0 \\ 1/2 & 0 & 0 & 3/4 & 0 & 1/4 \\ 0 & 0 & 0 & 0 & 1/4 & 0 \\ 1/2 & 0 & 0 & 1/4 & 0 & 1/2 \end{pmatrix}$$

where $\operatorname{diag}(w_1, \ldots, w_n)$ is the diagonal matrix with elements $w_1, \ldots, w_n$. If $n < \dim \mathscr{P}_2(\mathbb{R}^2) = 6$ then this matrix has rank < 6. But the matrix on the right-hand side is the so-called moment matrix, which is always non-singular and then has rank 6. Therefore, the above equality cannot be valid if $n < 6$.   □

A generalization of Theorem 2.2 is presented in Chap. 5.

Before giving the next lower bound, a certain class of integrals called *centrally symmetric integrals* is introduced.

Definition 2.2 (*Centrally symmetric integral*) An integral $\mathscr{I}$ is said to be *centrally symmetric* if $\mathscr{I}[f] = 0$ for all odd polynomials f.[2]

Example 2.4 Multivariate Gaussian integrals have been discussed in Example 2.2, i.e.,

---

[2] An odd polynomial f is a polynomial satisfying $f(-\omega) = -f(\omega)$ for all $\omega \in \Omega$.

$$\mathscr{I}[f] = \frac{1}{\pi^{d/2}} \int_{\mathbb{R}^d} f(\omega) e^{-\|\omega\|^2} \, d\omega$$

$$= \frac{1}{\pi^{d/2}} \int_{\mathbb{R}^d} f(\omega_1, \ldots, \omega_d) e^{-(\omega_1^2 + \cdots + \omega_d^2)} \, d\omega_1 \ldots d\omega_d.$$

This is a typical example of centrally symmetric integrals, which also belongs to a certain class of *spherically symmetric integrals*. Roughly speaking, $\mathscr{I}$ is called a spherically symmetric integral if region Ω has rotational invariance and the density function depends on the radial component of ω. The precise definition is given in the next section.

Theorem 2.2 is improved for odd-degree cubature formula for a centrally symmetric integral $\mathscr{I}$. The following lower bound is known as the *Möller bound*; see Möller [34, 35] and Mysovskikh [37].

Theorem 2.3 (Möller bound) *Let X be the set of points in a cubature formula of degree $2e + 1$ for a centrally symmetric integral $\mathscr{I}$. Then the following holds:*

$$|X| \geq \begin{cases} 2 \dim \mathscr{P}_e^*(\Omega) - 1 & \text{if } e \equiv 0 \pmod 2 \text{ and } 0 \in X, \\ 2 \dim \mathscr{P}_e^*(\Omega) & \text{otherwise.} \end{cases} \tag{2.2}$$

A sketch of proof of this theorem (along the lines given in Möller [34] or Stroud [58]) goes as follows.

Sketch of proof. Let $X = \{x_1, \ldots, x_n\}$ be an n-point subset of Ω with $0 \in X$, and let e be an even integer for simplicity. A key point of this proof is to consider functionals $L_x : \mathscr{P}_{2e+1}(\Omega) \to \mathbb{R}$ defined by

$$L_x(f) = f(x) \quad \text{for every } x \in X$$

as in (1.6).

First, it is known (e.g., [5, pp. 429–430]) that

$$\dim \operatorname{Span}_{\mathbb{R}} \{ L_{x_i} |_{\mathscr{P}_e^*(\Omega)} \mid i = 1, \ldots, n \} = \dim \mathscr{P}_e^*(\Omega).$$

Then, let $m = \dim \mathscr{P}_e^*(\Omega) - 1$ and $L_1, \ldots, L_m$ be functionals among $\{L_{x_1}, \ldots, L_{x_n}\} \setminus \{L_0\}$ such that $L_1 |_{\mathscr{P}_e^*(\Omega)}, \ldots, L_m |_{\mathscr{P}_e^*(\Omega)}$ is a basis of $\operatorname{Span}_{\mathbb{R}} \{ L_{x_i} |_{\mathscr{P}_e^*(\Omega)} \mid i = 1, \ldots, n \}$.

Since an odd polynomial $g \in \operatorname{Hom}_1(\Omega)$ with

$$L_0(g) = 0, \quad L_\ell(g) \neq 0, \quad \ell = 1, \ldots, m,$$

is obtainable, it can be checked that $L_1 |_{g \cdot \mathscr{P}_e^*(\Omega)}, \ldots, L_m |_{g \cdot \mathscr{P}_e^*(\Omega)}$ are linear independent. Moreover, since $g \cdot \mathscr{P}_e^*(\Omega)$ is a subset of $\mathscr{P}_{e+1}^*(\Omega)$, $L_1 |_{\mathscr{P}_{e+1}^*(\Omega)}, \ldots, L_m |_{\mathscr{P}_{e+1}^*(\Omega)}$ are also linear independent. This implies that

$$\dim \operatorname{Span}_{\mathbb{R}} \{ L_{x_i} |_{\mathscr{P}_e^*(\Omega)} \mid i = 1, \ldots, n \} - 1 = \dim \operatorname{Span}_{\mathbb{R}} \{ L_{x_i} |_{\mathscr{P}_{e+1}^*(\Omega)} \mid i = 1, \ldots, n \}.$$

Thus, it holds that

$$|X| \geq \dim \operatorname{Span}_{\mathbb{R}} \{ L_{x_i} |_{\mathscr{P}_{e+1}(\Omega)} \mid i = 1, \ldots, n \}$$
$$= \dim \operatorname{Span}_{\mathbb{R}} \{ L_{x_i} |_{\mathscr{P}_{e+1}^*(\Omega)} \mid i = 1, \ldots, n \} + \dim \operatorname{Span}_{\mathbb{R}} \{ L_{x_i} |_{\mathscr{P}_e^*(\Omega)} \mid i = 1, \ldots, n \}$$
$$= 2 \dim \operatorname{Span}_{\mathbb{R}} \{ L_{x_i} |_{\mathscr{P}_e^*(\Omega)} \mid i = 1, \ldots, n \} - 1,$$

which complete the proof. □

The following lemma is employed to explicitly calculate lower bound (2.1) or (2.2) for Ω, a set S_p of p concentric spheres centered at the origin; see also Example 2.5.

Lemma 2.1 (cf. [1]) *Let S_p be a set of p concentric spheres centered at the origin and $\varepsilon_{S_p} \in \{0, 1\}$ by*

$$\varepsilon_{S_p} = 1 \quad \text{if } 0 \in S_p, \quad \varepsilon_{S_p} = 0 \quad \text{if } 0 \notin S_p.$$

Then the following hold:

(i) *When $p \leq \lfloor \frac{t + \varepsilon_{S_p}}{2} \rfloor$*

$$\dim \mathscr{P}_t(S_p) = \varepsilon_{S_p} + \sum_{\ell=0}^{2(p - \varepsilon_{S_p}) - 1} \binom{d + t - \ell - 1}{t - \ell} < \binom{d + t}{t} = \dim \mathscr{P}_t(\mathbb{R}^d).$$

(ii) *When $p \geq \lfloor \frac{t + \varepsilon_{S_p}}{2} \rfloor + 1$*

$$\dim \mathscr{P}_t(S_p) = \sum_{\ell=0}^{t} \binom{d + t - \ell - 1}{t - \ell} = \binom{d + t}{t}.$$

(iii) *When $p \geq \lfloor t/2 \rfloor + 1$*

$$\dim \mathscr{P}_t^*(S_p) = \dim \mathscr{P}_t^*(\mathbb{R}^d) = \sum_{\ell=0}^{\lfloor t/2 \rfloor} \binom{d + t - 2\ell - 1}{t - 2\ell}.$$

(iv) *When $p \leq \lfloor t/2 \rfloor$ and t is odd or t is even and $0 \notin S_p$*

$$\dim \mathscr{P}_t^*(S_p) = \sum_{\ell=0}^{p-1} \binom{d + t - 2\ell - 1}{t - 2\ell} < \dim \mathscr{P}_t^*(\mathbb{R}^d).$$

(v) *When $p \leq \lfloor t/2 \rfloor$ with even t and $0 \in S_p$*

$$\dim \mathscr{P}_t^*(S_p) = 1 + \sum_{\ell=0}^{p-2} \binom{d + t - 2\ell - 1}{t - 2\ell} < \dim \mathscr{P}_t^*(\mathbb{R}^d).$$

(i)   **(ii)**

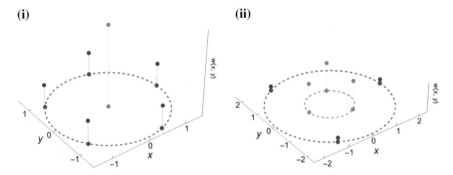

Fig. 2.3 Minimal cubature formulas of degree 4 of types (i) (left) and (ii) (right)

The definition of minimality of a cubature formula is now introduced.

Definition 2.3 (*Minimal cubature formula*) A cubature formula of degree t for $\mathscr{I}$ is said to be *minimal* if it attains (2.1) or (2.2), according as t is even or not.

Example 2.5 By using Lemma 2.1, bounds (2.1) and (2.2) are explicitly computed for cubature formulas of degrees 4 and 5 for the bivariate Gaussian integral as follows:

$$|X| \geq \dim \mathscr{P}_2(\mathbb{R}^2) = \binom{2+2}{2} = 6,$$

$$|X| \geq 2 \dim \mathscr{P}_2^*(\mathbb{R}^2) - 1 = 2\left\{\binom{3}{2} + \binom{1}{0}\right\} - 1 = 7.$$

Bannai et al. [6] shows that minimal cubature formulas of degree 4 are classified as the following two types (i) and (ii) (see also Fig. 2.3):
(i)

$$\mathscr{Q}[f] = \frac{1}{2}f(0,0) + \frac{1}{10}\sum_{\ell=0}^{4} f\left(\sqrt{2}\cos\frac{2\ell\pi}{5}, \sqrt{2}\sin\frac{2\ell\pi}{5}\right) \quad \text{for all } f \in \mathscr{P}_4(\mathbb{R}^2).$$

(ii)

$$\mathscr{Q}[f] = \frac{5+2\sqrt{5}}{30}\sum_{\ell=0}^{3} f\left(r_1\cos\frac{2\ell\pi}{3}, r_1\sin\frac{2\ell\pi}{3}\right)$$

$$+ \frac{5-2\sqrt{5}}{30}\sum_{\ell=0}^{3} f\left(r_2\cos\frac{(2\ell+1)\pi}{3}, r_2\sin\frac{(2\ell+1)\pi}{3}\right) \quad \text{for all } f \in \mathscr{P}_4(\mathbb{R}^2)$$

with $r_1 = (\sqrt{10} - \sqrt{2})/2$ and $r_2 = (\sqrt{10} + \sqrt{2})/2$.

Two illustrations in Fig. 2.3 may be thought of as a "discretization" of bivariate Gaussian distribution. For example, focusing on the left illustration, the weight at the origin (colored orange) indicates the peak of the density function and the other

Fig. 2.4 Minimal cubature
formula of degree 5 of type
(iii)

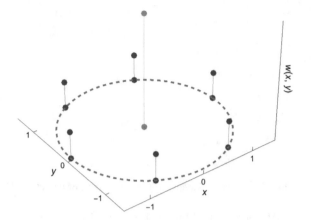

weights are identical and smaller than the peak at the remaining five points equidistant
from the origin (colored blue). Similarly for the right illustration, points on the
interlayer (colored orange) have identical weights which are larger than those on the
outer layer (colored blue).

Hirao and Sawa [28] also show that minimal cubature formulas of degree 5 are
classified as the following type (iii) (see also Fig. 2.4):

$$(\text{iii}) \quad \mathcal{Q}[f] = \frac{1}{2} f(0,0) + \frac{1}{12} \sum_{\ell=0}^{5} f\left(\sqrt{2}\cos\frac{\ell\pi}{3}, \sqrt{2}\sin\frac{\ell\pi}{3}\right) \quad \text{for all } f \in \mathcal{P}_5(\mathbb{R}^2).$$

In the remaining part of this section, a relationship between cubature formulas
and the (modified) kernel polynomials is discussed; see Sect. 1.1 for the definition
of (modified) kernel polynomials. The method of constructing cubature formulas
through reproducing kernels has been first considered by Mysovskikh [38, 39] and
later studied by Möller [33]; see also [14, 63] and references therein.

Given a positive integer e, let K_e be the (multivariate) kernel of $\mathcal{P}_e(\Omega)$ with
inner product $(f, g)_{\mathcal{P}_e(\Omega)}$. In order to construct a cubature formula of degree $2e$
for $\mathscr{I}$, this method needs to choose a d-point set $\{a_1, \ldots, a_d\}$ of $\mathbb{R}^d$ such that the
hypersurfaces $H_1, \ldots, H_d$ intersect at e^d points, where $H_\ell = \{\omega \in \mathbb{R}^d \mid K_e(\omega, a_\ell) =
0\}$, $\ell = 1, \ldots, e$. The following procedure explains how to choose $\{a_1, \ldots, a_d\}$ (see,
e.g., [14, 63]):

(i)   Pick any point a_1 in Ω which is not a common zero of $\{f_\ell\}_\ell$, where $\{f_\ell\}_\ell$ is an
      orthonormal basis of $\text{Hom}_e(\Omega)$.
(ii) Pick a point a_2 from the hypersurface H_1 which is not a common zero of $\{f_\ell\}_\ell$.
(iii) When the points $a_1, \ldots, a_k$ are selected, pick a point a_{k+1} from the intersection
      of hypersurfaces $H_1 \cap \ldots \cap H_k$ which is not a common zero of $\{f_\ell\}_\ell$.
(iv) Iterate (iii) until a_1 through a_d are selected.

Theorem 2.4 (cf. [39]) *If $H_1, \ldots, H_d$ defined by $a_1, \ldots, a_d$ intersect at e^d distinct
points, say $b_1, \ldots, b_{e^d}$, then there exists a cubature formula of degree $2e$ for $\mathscr{I}$ which
is of the form*

$$\mathcal{Q}[f] = \sum_{\ell=1}^{d} \frac{1}{K_e(a_\ell, a_\ell)} f(a_\ell) + \sum_{\ell=1}^{e^d} w_\ell f(b_\ell) \quad \text{for all } f \in \mathscr{P}_{2e}(\mathbb{R}^d), \qquad (2.3)$$

where w_ℓ are solutions of a certain system of linear equations determined by linear functional $\mathscr{I}[f] - \sum_\ell f(a_\ell)/K_e(a_\ell, a_\ell)$.

Moreover, when $\mathscr{I}$ has central symmetry, the following type of cubature formulas of degree $2e + 1$ may be constructed: Given a positive integer e, let $\tilde{K}_e$ be the modified kernel corresponding to $\mathscr{P}_e^*(\Omega)$ with inner product $(\cdot, \cdot)_{\mathscr{P}_e^*(\Omega)}$. Further let a d-point set $\{a_1, \ldots, a_d\}$ of $\mathbb{R}^d$ such that the hypersurfaces $\tilde{H}_1, \ldots, \tilde{H}_d$ intersect at e^d points, where $\tilde{H}_\ell = \{\omega \in \mathbb{R}^d \mid \tilde{K}_e(\omega, a_\ell) = 0\}$, $\ell = 1, \ldots, e$, and points a_i are chosen in the above-mentioned procedure with H_ℓ replaced by $\tilde{H}_\ell$.

Theorem 2.5 (cf. [39]) *Assume that $\mathscr{I}$ is a centrally symmetric integral. If $\tilde{H}_1, \ldots, \tilde{H}_d$ defined by $a_1, \ldots, a_d$ intersect at e^d points, say $b_1, \ldots, b_{e^d}$, then there exists a cubature formula of degree $2e + 1$ for $\mathscr{I}$ which is of the form*

$$\mathcal{Q}[f] = \sum_{\ell=1}^{d} \frac{1}{2\tilde{K}_e(a_\ell, a_\ell)} \{f(a_\ell) + f(-a_\ell)\} + \sum_{\ell=1}^{e^d} w_\ell f(b_\ell) \quad \text{for all } f \in \mathscr{P}_{2e+1}(\mathbb{R}^d),$$

$$(2.4)$$

where w_ℓ are solutions of a certain system of linear equations determined by linear functional $\mathscr{I}[f] - \sum_{\ell=1}^{d} (f(a_\ell) + f(-a_\ell))/2\tilde{K}_e(a_\ell, a_\ell)$.[3]

Example 2.6 ([63]) A cubature formula of degree 5 for $\int_{\mathbb{B}^2} \cdot\, d\omega / \int_{\mathbb{B}^2} d\omega$ is constructed as follows. Set $a_1 = (0, 0)$. Since $\tilde{K}_2(\omega, \omega') := \tilde{K}_2^{(1/2)}(\omega, \omega') = 12\langle\omega, \omega'\rangle^2 + 6\|\omega\|^2\|\omega'\|^2 - 6(\|\omega\|^2 + \|\omega'\|^2) + 4$ from Theorem 1.6, it holds that

$$\tilde{H}_1 = \{\omega \in \mathbb{R}^2 \mid \tilde{K}_2(\omega, a_1) = 0\} = \{\omega \in \mathbb{R}^2 \mid \|\omega\|^2 = 2/3\}.$$

Let $a_2 = (2/\sqrt{6}, 0)$. Since

$$\tilde{H}_2 = \{\omega \in \mathbb{R}^2 \mid \tilde{K}_2(\omega, a_2) = 0\} = \{(\omega_1, \omega_2) \in \mathbb{R}^2 \mid 6\omega_1^2 - 2\omega_2^2 = 0\},$$

it is easy to check that $\{b_1, \ldots, b_4\} = \{(\pm 1/\sqrt{6}, \pm 1/\sqrt{2})\} \in H_1 \cap H_2$. Thus, it follows from Theorem 2.5 that

$$\mathcal{Q}[f] = \frac{1}{4} f(0, 0) + \frac{1}{8} \left\{ f\left(\frac{2}{\sqrt{6}}, 0\right) + f\left(-\frac{2}{\sqrt{6}}, 0\right) \right\} + \frac{1}{8} \sum f\left(\pm\frac{1}{\sqrt{6}}, \pm\frac{1}{\sqrt{2}}\right)$$

$$= \frac{1}{4} f(0, 0) + \frac{1}{8} \sum_{\ell=0}^{5} f\left(\frac{\sqrt{6}}{3} \cos\frac{\ell\pi}{3}, \frac{\sqrt{6}}{3} \sin\frac{\ell\pi}{3}\right) \quad \text{for all } f \in \mathscr{P}_5(\mathbb{R}^2),$$

which forms a minimal cubature formula of degree 5; see also Example 2.5.

---

[3] See [39, Theorems 12.1,12.2] for details of the proofs of Theorems 2.4 and 2.5.

The method of Theorem 2.4 (respectively, Theorem 2.5) is based on reproducing kernels K_e (respectively, $\tilde{K}_e$), which is of theoretical interest from a viewpoint of functional analysis. We refer the reader to Xu [63] who explains some merits of Theorems 2.4 and 2.5 are explained. In some special cases, K_e and $\tilde{K}_e$ have a quite elegant form that represents (2.3) and (2.4) in a more explicit manner. For example, Xu [63] uses a compact formula of K_e and $\tilde{K}_e$ for multivariate Jacobi integral (see Theorem 1.6 in Sect. 1.2) and thereby produces many examples of cubature formulas.

An important application of reproducing kernels to the cubature theory is to characterize minimal cubature formulas, which can be regarded as a high-dimensional generalization of Gaussian quadrature formula.

Theorem 2.6 ([36, 39])

(1) *Assume that there exists a cubature formula of degree $2e$ for an integral $\mathscr{I}$. This formula is minimal if and only if*

> *(i) $K_e(x, x') = 0$ for all $x, x' \in X$ with $x \neq x'$,*
> *(ii) $w(x) = 1/K_e(x, x)$ for all $x \in X$.*

(2) *Assume that there exists a cubature formula of degree $4k + 1$ for a centrally symmetric integral $\mathscr{I}$ of the form*

$$\mathscr{Q}[f] = w(0)f(0) + \sum_{x \in \tilde{X} \setminus \{0\}} w(x)\{f(x) + f(-x)\}.$$

This formula is minimal if and only if

> *(i) $\tilde{K}_{2k}(x, x') = 0$ for all $x, x' \in X$ with $x \neq \pm x'$,*
> *(ii) $w(0) = 1/\tilde{K}_{2k}(0, 0)$, $w(x) = 1/(2\tilde{K}_{2k}(x, x))$ for all $x \in X \setminus \{0\}$.*

(3) *Assume that there exists a cubature formula of degree $4k + 3$ for a centrally symmetric integral $\mathscr{I}$ of the form*

$$\mathscr{Q}[f] = \sum_{x \in \tilde{X}} w(x)\{f(x) + f(-x)\}.$$

This formula is minimal if and only if

> *(i) $\tilde{K}_{2k+1}(x, x') = 0$ for all $x, x' \in X$ with $x \neq \pm x'$,*
> *(ii) $w(x) = 1/(2\tilde{K}_{2k+1}(x, x))$ for all $x \in X$.*

Example 2.7 By using Theorem 2.6 (2), Noskov and Schmid [43] give a classification of minimal cubature formulas of degree 5 for $\int_{\mathbb{B}^d} \cdot\, d\omega / \int_{\mathbb{B}^d} d\omega$. As a by-product, they show that

$$X = \{(0, 0, 0, 0, 0, 0, 0), \tfrac{3}{\sqrt{11}}(\pm 1, \pm 1, 0, \pm 1, 0, 0, 0),$$
$$\tfrac{3}{\sqrt{11}}(0, \pm 1, \pm 1, 0, \pm 1, 0, 0), \tfrac{3}{\sqrt{11}}(0, 0, \pm 1, \pm 1, 0, \pm 1, 0),$$
$$\tfrac{3}{\sqrt{11}}(0, 0, 0, \pm 1, \pm 1, 0, \pm 1), \tfrac{3}{\sqrt{11}}(\pm 1, 0, 0, 0, \pm 1, \pm 1, 0),$$
$$\tfrac{3}{\sqrt{11}}(0, \pm 1, 0, 0, 0, \pm 1, \pm 1), \tfrac{3}{\sqrt{11}}(\pm 1, 0, \pm 1, 0, 0, 0, \pm 1)\}$$

forms a 57-point minimal cubature formula of degree 5 for $\int_{\mathbb{B}^d} \cdot\, d\omega / \int_{\mathbb{B}^d} d\omega$, which is the same structure as the 57-point minimal cubature for $\int_{\mathbb{R}^d} \cdot \exp(-\|\omega\|^2)\, d\omega / \pi^{d/2}$ found by Victoir [62]. Note that $X \setminus \{(0, \dots, 0)\}$ has the same arrangement as Box–Behnken design No. 5 in [11, Table 4]; see also Example 4.8 in Sect. 4.4. Moreover, $X \setminus \{(0, \dots, 0)\}$ is essentially the same as the characteristic functions (vectors) of blocks of a 2-(7, 3, 1)-design, which is the smallest nontrivial example of combinatorial 2-designs, i.e., balanced incomplete block (BIB) designs. See Sect. 4.4 for the definition of combinatorial designs.

Hirao and Sawa [27] extend the Noskov–Schmid formula to minimal cubature formulas for spherically symmetric integrals in general; see Definition 2.7 for the definition of spherical symmetry.

The next section discusses the relationship between cubature formulas and Euclidean designs, which will be used for subsequent Chaps. 3 and 4.

2.3 Euclidean Design and Spherical Symmetry

The present section first gives the concept of *spherical design* defined by Delsarte et al. [17], which is one of main research objects in algebraic combinatorics.

Definition 2.4 (*Spherical design*) Given a nonnegative integer t, a finite subset X of $\mathbb{S}^{d-1}$ is called a *spherical t-design* if

$$\frac{1}{|\mathbb{S}^{d-1}|} \int_{\mathbb{S}^{d-1}} f(\omega)\, \rho(d\omega) = \frac{1}{|X|} \sum_{x \in X} f(x) \quad \text{for all } f \in \mathscr{P}_t(\mathbb{R}^d). \tag{2.5}$$

As immediately seen by Definition 2.4, a spherical t-design of $\mathbb{S}^{d-1}$ is just a cubature formula of degree t for spherical integration. In analysis language, this is just an equal-weighted cubature formula, called a *Chebyshev-type cubature formula* (e.g., [57]). The Chebyshev-type cubature shows an extension of the famous Chebyshev–Gauss quadrature as

$$\int_{-1}^{1} f(u) \frac{du}{\pi\sqrt{1 - u^2}} = \frac{1}{e} \sum_{\ell=1}^{e} f\left(\cos \frac{2\ell - 1}{2e}\pi\right) \quad \text{for all } f \in \mathscr{P}_{2e-1}(\mathbb{R}),$$

where $\cos\left((2\ell - 1)\pi/2e\right)$ are the zeros of the first kind Chebyshev polynomial of degree e. This is minimal with respect to (2.3) among all quadrature formulas of degree $2e - 1$ for $\int_{-1}^{1} \cdot 1/(\pi\sqrt{1 - u^2})\, du$.

Remark 2.1 Neumaier and Seidel [41] show the equivalence between optimal experimental designs of degree e on the unit sphere and spherical $2e$-designs where X may be a multi-set. This relationship will be explained in detail in Sect. 3.2. Moreover, a spherical design with rational weights and rational points is closely connected with isometric embeddings of classical finite-dimensional Banach spaces or a certain class of polynomial identities called Hilbert identities; for details see Sects. 2.4 and 5.6.

Remark 2.2 Polar coordinate systems are used when calculating a spherical integration. Polar coordinates $r, \theta_1, \ldots, \theta_{d-1}$ are converted to $\omega_1, \ldots, \omega_d$ by using trigonometric functions as

$$\omega_1 = r\sin\theta_{d-1}\sin\theta_{d-2}\cdots\sin\theta_2\sin\theta_1$$
$$\omega_2 = r\sin\theta_{d-1}\sin\theta_{d-2}\cdots\sin\theta_2\cos\theta_1$$
$$\vdots$$
$$\omega_i = r\sin\theta_{d-1}\sin\theta_{d-2}\cdots\sin\theta_i\cos\theta_{i-1}$$
$$\vdots$$
$$\omega_{d-1} = r\sin\theta_{d-1}\cos\theta_{d-2}$$
$$\omega_d = r\cos\theta_{d-1},$$

where $0 \le r, 0 \le \theta_1 < 2\pi$ and $0 \le \theta_i \le \pi$ for $i = 2, \ldots, d - 1$. Moreover, the corresponding volume element is given by $r^{d-1}\prod_{\ell=2}^{d-1}\sin^{\ell-1}\theta_\ell\, d\theta_{d-1}d\theta_{d-2}\cdots d\theta_1$. By using this system, the surface area of $\mathbb{S}^{d-1}$ is calculated as follows:

$$|\mathbb{S}^{d-1}| = \int_{\mathbb{S}^{d-1}} \rho(d\omega)$$

$$= \int_0^{2\pi}\int_0^\pi \cdots \int_0^\pi \left(\prod_{\ell=2}^{d-1}\sin^{\ell-1}\theta_\ell\right) d\theta_{d-1}\cdots d\theta_1$$

$$= \int_0^{2\pi}\int_0^\pi \cdots \int_0^\pi \left(\prod_{\ell=2}^{d-2}\sin^{\ell-1}\theta_\ell\right) d\theta_{d-2}\cdots d\theta_1 \cdot \int_0^\pi \sin^{d-2}\theta_{d-1}\, d\theta_{d-1}$$

$$= |\mathbb{S}^{d-2}| \cdot \int_0^\pi \sin^{d-2}\theta_{d-1}\, d\theta_{d-1}$$

$$= |\mathbb{S}^{d-2}| \cdot \frac{\pi^{1/2}\Gamma((d-1)/2)}{\Gamma(d/2)},$$

where Γ is the Gamma function. By using the abovementioned procedure repeatedly, it holds that $|\mathbb{S}^{d-1}| = 2\pi^{d/2}/\Gamma(d/2)$.

Moreover, with the multi-index notation $\omega^a = \omega_1^{a_1} \cdots \omega_d^{a_d}$ for $a = (a_1, \ldots, a_d) \in \mathbb{Z}_{\geq 0}^d$, it is known (e.g., [22]) that

$$
\frac{1}{|\mathbb{S}^{d-1}|} \int_{\mathbb{S}^{d-1}} \omega^a \, \rho(d\omega) =
\begin{cases}
\dfrac{\Gamma(d/2)}{2^{\|a\|_1} \Gamma(\|a\|_1/2 + d/2)} \cdot \dfrac{\prod_{j=1}^d (a_j)!}{\prod_{j=1}^d (a_j/2)!}, & \text{if } a_j \text{ are even for all } j; \\
0, & \text{otherwise}
\end{cases}
$$

(2.6)

with $\|a\|_1 = a_1 + \cdots + a_d$, which will be used in the proof of Theorem 5.1.

Now, a fundamental question concerns the existence of spherical t-designs. Seymour and Zaslavsky [53] show that there exists a spherical t-design for sufficiently large number of points (see also Corollary 5.2). A recent work by Bondarenko et al. [9] is an improvement of Seymour–Zaslavsky theorem ([53]).

Theorem 2.7 (Bondarenko–Radchenko–Viazovska theorem, [9]) *Given $d \geq 2$, there exists a spherical t-design on $\mathbb{S}^{d-1}$ with $|X|$ points for all $|X| \geq c_d t^{d-1}$ where the constant c_d does only depend on d.*

Note that this lower bound has the same asymptotic behavior as the Fisher-type bound (2.1) for fixed d. Moreover, Bondarenko et al. [10] show a similar existence result for "well-separated" spherical designs X, i.e., X satisfies $|x_i - x_j| \geq c_d' n^{-1/(d-1)}$, $i \neq j$, where the constant c_d' does only depend on d.

Next, the tightness of spherical designs is defined. In algebraic combinatorics, tight designs are of particular interest for the mathematical richness of its own structure.

Definition 2.5 (*Tight spherical design*) A spherical t-design is said to be *tight* if the equality in (2.1) or (2.2) is attained for $\Omega = \mathbb{S}^{d-1}$, according as t is even or not.

Remark 2.3 In numerical analysis, *minimal* is often used for *tight*; recall Definition 2.3.

Example 2.8 (*Tight spherical designs of* $\mathbb{S}^1$) Given a positive integer t, the vertices of a regular $(t + 1)$-gon inscribed in $\mathbb{S}^1$ form a tight spherical t-design; see, e.g., [2, 3]. For example, the vertices of a regular pentagon inscribed in $\mathbb{S}^1$ form a tight spherical 4-design, i.e., it holds that

$$
\frac{1}{2\pi} \int_0^{2\pi} f(\cos\theta, \sin\theta) \, d\theta = \frac{1}{5} \sum_{\ell=0}^4 f\left(\cos\frac{2\ell\pi}{5}, \sin\frac{2\ell\pi}{5}\right) \quad \text{for all } f \in \mathscr{P}_4(\mathbb{R}^2).
$$

While tight spherical designs always exist on $\mathbb{S}^1$, the situation becomes quite different in higher dimensional cases, i.e., tight designs only sporadically exist in higher dimensions. The known results on the classification on tight spherical t-designs of $\mathbb{S}^{d-1}$ ($d \geq 3$) are summarized as follows ([2, 3]):

(i) X is a tight spherical 1-design of $\mathbb{S}^{d-1}$ if and only if $X = \{x, -x\}$ is an antipodal set with $x \in \mathbb{S}^{d-1}$.

(ii) X is a tight spherical 2-design of $\mathbb{S}^{d-1}$ if and only if X is a regular simplex.

(iii) X is a tight spherical 3-design of $\mathbb{S}^{d-1}$ if and only if X is isomorphic to a cross polytope $\{\pm e_\ell \mid 1 \leq \ell \leq d\}$, where $e_1, \ldots, e_d$ is the standard basis of $\mathbb{R}^d$.

(iv) There does not exist a tight spherical t-design of $\mathbb{S}^{d-1}$ with possible exceptions that $t \leq 5$ and $t \in \{7, 11\}$.

 (v) The classifications of tight spherical 4-, 5-, 7-designs are not complete yet; see also Bannai et al. [7]. As of today, the only known examples are as follows:

   a. $t = 4$: a 27-point set on $\mathbb{S}^5$ related to the E_6 root system, a 275-point set on $\mathbb{S}^{21}$ related to the McLaughlin graph.
   b. $t = 5$: a set of 12 vertices of the icosahedron on $\mathbb{S}^2$, a set of 126 vectors of the E_7 root system on $\mathbb{S}^6$, a 552-point set on $\mathbb{S}^{22}$ related to 276 equiangular lines.
   c. $t = 7$: a set of 240 vectors of the E_8 root system on $\mathbb{S}^7$, a 4600-point set on $\mathbb{S}^{23}$ related to an iterated kissing configuration.

(vi) X is a tight spherical 11-design of $\mathbb{S}^{23}$ if and only if X is isomorphic to a set of 196560 minimal vectors in the Leech lattice in $\mathbb{R}^{24}$.

Neumaier and Seidel [40] give the concept of *Euclidean designs* as a generalization of spherical designs; see also Delsarte and Seidel [18].

Let X be a finite subset of $\mathbb{R}^d$ and w be a positive weight function on X. Denote by $\mathscr{R}$ the radius set of X, i.e.,

$$\mathscr{R} = \{\|x\| \mid x \in X\} = \{r_1, \ldots, r_p\} \text{ with } r_1 > r_2 > \cdots > r_p \geq 0.$$

Let $\mathbb{S}_{r_i}^{d-1} = \{\omega \in \mathbb{R}^d \mid \|\omega\| = r_i\}$ be the $(d-1)$-dimensional sphere with radius r_i and $S_p = \cup_{i=1}^p \mathbb{S}_{r_i}^{d-1}$ be a set of p concentric spheres centered at the origin. Further, let $X_i = X \cap \mathbb{S}_{r_i}^{d-1}$ and $W_i = \sum_{x \in X_i} w(x)$. Let ρ_{r_i} be the surface measure and $|\mathbb{S}_{r_i}^{d-1}|$ be the surface area of $\mathbb{S}_{r_i}^{d-1}$. For notational convenience, when $r_p = 0$, we let $\int_{\mathbb{S}_{r_p}^{d-1}} f(\omega) \, \rho_{r_p}(d\omega)/|\mathbb{S}_{r_p}^{d-1}| = f(0)$. Then, consider a cubature formula for the integral $\mathscr{I}'$ defined as

$$\mathscr{I}'(f) = \sum_{i=1}^p \frac{W_i}{|\mathbb{S}_{r_i}^{d-1}|} \int_{\mathbb{S}_{r_i}^{d-1}} f(\omega) \, \rho_{r_i}(d\omega).$$

Under this setting, the following definition is introduced.

Definition 2.6 (*Euclidean design*) Given a nonnegative integer t, a finite weighted pair (X, w) is called a *Euclidean t-design* supported by a set S_p of p concentric spheres if

$$\mathscr{I}'[f] = \mathscr{Q}[f] \quad \text{for all } f \in \mathscr{P}_t(\mathbb{R}^d). \tag{2.7}$$

Remark 2.4 (i) If $p = 1$, $r_1 = 1$ and $w \equiv 1$ on X, then X is a spherical t-design of $\mathbb{S}^{d-1}$ (see Definition 2.4).

(ii) If $0 \in X$, then $X \setminus \{0\}$ is also a Euclidean t-design.

The following theorem is known as the *Neumaier–Seidel equivalence theorem*.

Theorem 2.8 (Neumaier–Seidel equivalence theorem, [40]) *Let (X, w) be a finite weighted pair. Then the following are equivalent:*

(i) A pair (X, w) is a Euclidean t-design.
(ii) $\sum\limits_{x \in X} w(x)\|x\|^{2j}\phi(x) = 0$ for every $\phi \in \mathrm{Harm}_\ell(\mathbb{R}^d)$, with $1 \le \ell \le t$, $0 \le j \le \lfloor\frac{t-\ell}{2}\rfloor$.

Proof It is known (e.g., [40]) that the polynomial space $\mathscr{P}_t(\mathbb{R}^d)$ is decomposed as a finite sum of products of $\|\omega\|^{2j}$ and harmonic homogeneous polynomial spaces $\mathrm{Harm}_\ell(\mathbb{R}^d)$ with $0 \le 2j + \ell \le t$, i.e.,

$$\mathscr{P}_t(\mathbb{R}^d) = \bigoplus_{0 \le 2j+\ell \le t} \|\omega\|^{2j}\,\mathrm{Harm}_\ell(\mathbb{R}^d).$$

This is the $\mathbb{R}^d$-version of the Fischer decomposition (1.13). This implies the desired result. □

An algebraic generalization of this result will be available in Sect. 4.2.

Let us look at the relationship between Euclidean designs and cubature formulas for spherically symmetric integrals.

Example 2.9 A multivariate Gaussian integral, which has appeared several times in this chapter, i.e.,

$$\mathscr{I}[f] = \frac{1}{\pi^{d/2}} \int_{\mathbb{R}^d} f(\omega)e^{-\|\omega\|^2}\, d\omega$$
$$= \frac{1}{\pi^{d/2}} \int_{\mathbb{R}^d} f(\omega_1, \dots, \omega_d)e^{-(\omega_1^2 + \cdots + \omega_d^2)}\, d\omega_1 \dots d\omega_d$$

is a typical example of spherically symmetric integrals. Roughly speaking, if the region Ω has "rotational invariance" and the density function depends on the radial component of ω, then $\mathscr{I}$ is a spherically symmetric integral. Particularly, in the 2-dimensional case, the term *circularly symmetric integral* or *radially symmetric integral* is often used.

Let $\Omega = \{\omega = (\omega_1, \dots, \omega_d) \in \mathbb{R}^d \mid \alpha \le \|\omega\| \le \beta\}$ be a connected annulus region with nonnegative real values α, β with $\alpha < \beta$, and W be a radial weight function, i.e., a density function of $\|\omega\|$ for all $\omega \in \Omega$. Under this setting, the following definition is introduced.

Definition 2.7 (*Spherically symmetric integral*) With the above setting, integral $\mathscr{I}$ is said to be *spherically symmetric* if it is of the form

$$\mathscr{I}[f] = \frac{1}{\int_\Omega W(\|\omega\|)\,d\omega} \int_\Omega f(\omega) W(\|\omega\|)\,d\omega$$

$$= \frac{1}{\int_\alpha^\beta W(r)r^{d-1}\,dr} \int_\alpha^\beta \left(\frac{1}{|\mathbb{S}^{d-1}|} \int_{\mathbb{S}^{d-1}} f(r\omega)\,\rho(d\omega) \right) r^{d-1} W(r)\,dr.$$

The Neumaier–Seidel equivalence theorem (Theorem 2.8) gives the close relationship between Euclidean designs and cubature formulas for spherically symmetric integrals $\mathscr{I}$.

Corollary 2.1 *A cubature formula of degree t for a spherically symmetric integral $\mathscr{I}$ forms a Euclidean t-design.*

Proof For $\ell \geq 1$ and $\phi \in \mathrm{Harm}_\ell(\mathbb{R}^d)$,

$$\sum_{x \in X} w(x)\|x\|^{2j}\phi(x) = \frac{1}{\int_\Omega W(\|\omega\|)\,d\omega} \int_\Omega \|\omega\|^{2j}\phi(\omega) W(\|\omega\|)\,d\omega$$

$$= \frac{1}{\int_\Omega W(\|\omega\|)\,d\omega} \int_\alpha^\beta r^{\ell+2j+d-1} W(r)\,dr \int_{\mathbb{S}^{d-1}} \phi(\omega)\,\rho(d\omega)$$

$$= 0.$$

Thus, the desired result is obtained from Theorem 2.8. $\square$

Definition 2.8 (*Tight Euclidean design*) A Euclidean t-design is said to be *tight* if the equality in (2.1) or (2.2) is attained for $\Omega = S_p$, according as t is even or not.

Example 2.10 By Corollary 2.1, the examples given in Example 2.5 are tight Euclidean designs. There exist some classification results of tight t-designs for small values of t. For example, Bannai [1] gives a classification of tight Euclidean 5-designs in low dimensions and presents some new examples of tight 5-designs, namely, let $X = X_1 \cup X_2$ be a finite subset of $\mathbb{R}^6$, where

$$X_1 = \{\pm e_i \mid i = 1, \ldots, 6\}, \quad X_2 = \left\{ \frac{r}{\sqrt{6}} \mid \varepsilon_i \in \{1, -1\}, |\{i \mid \varepsilon_i = 1\}| \equiv 0 \pmod 2 \right\}$$

and w is a positive weight function such that $w(x) = 1$ for $x \in X_1$ and $w(x) = 9/(8r^4)$ for $x \in X_2$. Then a pair (X, w) forms a tight Euclidean 5-design on two concentric spheres. Note that this tight design is invariant under reflection group D_6; for details on reflection groups, see Chap. 4.

A technique for constructing Euclidean designs is presented. The idea is almost trivial. However, it is of great important from a viewpoint of design of experiments in connection with optimal design theory; see, e.g., Theorem 4.10.

Proposition 2.1 *Let (Y_i, w_i), $i = 1, \ldots, p$, be Euclidean t-designs on the unit sphere. Let $\Lambda_1, \ldots, \Lambda_p$ and $r_1, \ldots, r_p$ be positive real numbers. Then $(\cup_{i=1}^p r_i Y_i, w)$ is a Euclidean t-design on $S_p = \cup_{i=1}^p \mathbb{S}_{r_i}^{d-1}$, where $w(\omega) = \Lambda_i w_i(r_i^{-1}\omega)$ if $r_i^{-1}\omega \in Y_i$.*

Proof The polynomial space $\mathscr{P}_t(\mathbb{R}^d)$ is decomposed as a finite sum of homogeneous polynomial spaces $\mathrm{Hom}_\ell(\mathbb{R}^d)$ with $0 \le \ell \le t$; see, e.g., Neumaier and Seidel [41]. Given positive real numbers $\Lambda_1, \ldots, \Lambda_p$, it holds that for all $f \in \mathrm{Hom}_\ell(\mathbb{R}^n)$ with $0 \le \ell \le t$,

$$\sum_{i=1}^p \frac{\Lambda_i}{|\mathbb{S}_{r_i}^{d-1}|} \int_{\mathbb{S}_{r_i}^{d-1}} f(\omega)\, \rho_{r_i}(d\omega) = \sum_{i=1}^p \frac{r_i^\ell \Lambda_i}{|\mathbb{S}^{d-1}|} \int_{\mathbb{S}^{d-1}} f(\omega)\, \rho(d\omega)$$

$$= \sum_{i=1}^p r_i^\ell \Lambda_i \sum_{x \in Y_i} w_i(x) f(x)$$

$$= \sum_{i=1}^p \sum_{x \in Y_i} (\Lambda_i w_i(x)) f(r_i x),$$

which implies the desired result. $\qquad\qquad\square$

Moreover, the following is a simple but not entirely trivial construction of cubature formulas for spherically symmetric integrals.

Proposition 2.2 (cf. [26]) *Assume that there exist a Euclidean t-design* (Y, w) *on the unit sphere and a quadrature formula* $(\{r_i^2\}_i, \{\Lambda_i\}_i)$ *of degree* $\lfloor t/2 \rfloor$ *for integral*

$$\mathscr{I}[f] = \frac{1}{\int_{\alpha^2}^{\beta^2} u^{d/2-1} W(\sqrt{u})\, du} \int_{\alpha^2}^{\beta^2} f(u) u^{d/2-1} W(\sqrt{u})\, du.$$

Then a finite weighted pair $(\{r_i Y\}_i, \{\Lambda_i w\}_i)$ *forms a cubature formula of degree t for a spherically symmetric integral* $\mathscr{I}$.

Proof By recalling the argument used in the proof of Theorem 2.8, polynomial space $\mathscr{P}_t(\mathbb{R}^d)$ is decomposed as

$$\mathscr{P}_t(\mathbb{R}^d) = \bigoplus_{0 \le 2j+\ell \le t} \|\omega\|^{2j} \mathrm{Harm}_\ell(\mathbb{R}^d).$$

For all $f \in \|\omega\|^{2j} \mathrm{Harm}_l(\mathbb{R}^d)$ with $1 \le \ell \le t$ and $0 \le j \le \lfloor (t-\ell)/2 \rfloor$, it holds that

$$\sum_{i=1}^p \sum_{x \in Y} (\Lambda_i w(x)) f(r_i x) = \sum_{i=1}^p r_i^{\ell+2j} \Lambda_i \sum_{x \in Y} w(x) f(x)$$

$$= \sum_{i=1}^p r_i^{\ell+2j} \frac{\Lambda_i}{|\mathbb{S}^{d-1}|} \int_{\mathbb{S}^{d-1}} f(\omega)\, \rho(d\omega)$$

$$= \sum_{i=1}^p \frac{\Lambda_i}{|\mathbb{S}_{r_i}^{d-1}|} \int_{\mathbb{S}_{r_i}^{d-1}} f(\omega)\, \rho_{r_i}(d\omega) = 0. \qquad (2.8)$$

Moreover, for all $f = \|\omega\|^{2j}$ with $j = 0, 1, \ldots, \lfloor t/2 \rfloor$, it holds that

$$\sum_{i=1}^{p} \Lambda_i r_i^{2j} = \frac{1}{\int_{\alpha^2}^{\beta^2} u^{d/2-1} W(\sqrt{u})\, du} \int_{\alpha^2}^{\beta^2} u^j \cdot u^{d/2-1} W(\sqrt{u})\, du. \qquad (2.9)$$

Combining (2.8) and (2.9) gives the desired result.   □

The existence of Gaussian quadrature (e.g., [58]) always implies the existence of a quadrature formula of degree t for $\mathcal{J}$ with $\lfloor t/2 \rfloor$ points. Therefore, if a spherical design can be explicitly constructed, then so is a cubature formula of degree t for a spherically symmetric integral $\mathcal{I}$.

Note that these observations will be used in Proposition 3.1.

2.4   Further Remarks and Open Questions

In Sect. 2.2, two classical lower bounds, the Fisher-type bound (Theorem 2.2) and the Möller bound (Theorem 2.3), have been derived by using standard techniques in linear algebra and functional analysis. Delsarte et al. [17] show the same lower bounds for spherical designs in a unified approach based on the linear programming (LP) techniques. A sketch of their idea is as follows: Given positive integers d and t, let $\mathscr{F}_{d,t}$ be a collection of continuous functions f on interval $[-1, 1]$ such that

(i) $f(u) \geq 0$ for all $u \in [-1, 1]$ and $f(1) > 0$,
(ii) $f_\ell \geq 0$ for all $\ell \geq t + 1$, where

$$f_\ell = \frac{\int_{-1}^{1} f(u) C_\ell^{((d-2)/2)}(u)(1 - u^2)^{(d-3)/2}\, du}{\int_{-1}^{1} \{C_\ell^{((d-2)/2)}(u)\}^2 (1 - u^2)^{(d-3)/2}\, du}.$$

Here $C_l^{((d-2)/2)}$ denotes the Gegenbauer polynomial defined by (1.8) in Sect. 1.2. Then the number of points in a spherical t-design X is bounded from below as

$$|X| \geq \sup_{f \in \mathscr{F}_{d,t}} \frac{f(1)}{f_0}.$$

To derive the Fisher-type bound (Theorem 2.2) and the Möller bound (Theorem 2.3), Delsarte et al. [17] choose f as follows:

$$f(u) = \begin{cases} \left(\displaystyle\sum_{\ell=0}^{e} Q_\ell(u) \right)^2, & \text{if } t = 2e, \\[4ex] (u + 1) \left(\displaystyle\sum_{\ell=0}^{\lfloor e/2 \rfloor} Q_{e-2\ell}(u) \right)^2, & \text{if } t = 2e + 1. \end{cases} \qquad (2.10)$$

With the symbols Q_ℓ (in Remark 1.4) and P_ℓ and R_ℓ (in Example 1.7), (2.10) is simplified as $(R_e(u))^2$ and $(u+1)(P_e(u))^2$. Direct calculations show that $f_0 = R_e(1)$ if $t = 2e$, and $f_0 = P_e(1)$ if $t = 2e + 1$. It thus remains to calculate $R_e(1)$ and $P_e(1)$. For example, when $t = 2e + 1$, recall (1.16), i.e., the computation of $\dim \mathrm{Hom}_e(\mathbb{S}^{d-1})$, which is just the computation of $P_e(1)$. As implied by (1.17), the computation of $R_e(1)$ is substantially the same as that of $\dim \mathscr{P}_e(\mathbb{S}^{d-1})$. In summary, bounds (2.1) and (2.2) are obtained in a unified fashion. For an improvement of the Delsarte LP bound, we refer the reader to Yudin [65].

As mentioned in Sect. 1.2, Xu [64] finds an analogue of the Delsarte LP bound for multivariate Jacobi integral over unit ball $\mathbb{B}^d$ by combining the above techniques with "compact" expression of reproducing kernels (Theorem 1.6). A challenging problem is to find such compact expressions of the kernel function corresponding to Gaussian integration by using classical orthogonal polynomials like Xu's compact formula (Theorem 1.6). Gaussian integration may be of particular interest from a viewpoint of the cubature theory and statistical applications (e.g., [16, 31]). Not only is the LP bound improved for Gaussian integration as mentioned in the unit ball case, but also many new cubature formulas can be generated by the Mysovskikh theorems (Theorems 2.4 and 2.5).

In the context of *space-filling design* (e.g., [45]), a numerical integration method called the *quasi-Monte Carlo (QMC) method* on unit cube $\Omega = [0, 1]^d$ is often used. The aim of QMC method on a RKHS $\mathscr{H}$ is to construct sequences $\{X_n\}_n$, where each X_n is an n-point configuration on Ω, which gives a certain optimal convergence rate, faster than the Monte Carlo rate of $O(n^{-1/2})$, for the worst-case error defined as

$$\mathrm{wce}(\mathscr{H}; X_n) := \sup_{f \in \mathscr{H}, \|f\|_{\mathscr{H}} \le 1} \left| \int_\Omega f(\omega)\, d\omega - \frac{1}{n} \sum_{\ell=1}^n f(x_\ell) \right|. \tag{2.11}$$

The reproducing kernel theory discussed in Chap. 1 plays a key role in investigating a convergent rate of (2.11) for given sequences of point configurations. Then a number of "good" sequences, such as the Sobol sequence, lattice rules, digital nets, and so on, have been proposed; see Dick and Pillichshammer [19], Niederreiter [42] and references therein.

On the other hand, works on spherical cases $\Omega = \mathbb{S}^{d-1}$ have been increasing in recent years. Brauchart et al. [12] introduce the concept of QMC design sequences for some RKHS. For example, they show that an increasing sequence of n-point spherical t-designs X_n on $\mathbb{S}^{d-1}$ with $n = O(t^{d-1})$ forms a *QMC design sequence*.

Comparison of numerical accuracy of quasi-Monte Carlo approaches and cubature approaches is discussed by several authors. Schürer [51, 52] performs numerical experiments with Genz functions [23] by choosing classical QMC points such as Sobol sequences, Niederreiter sequences (cf. [19]) and some classes of classical cubatures, as exemplified by Stroud formula (see [58]), in dimension up to 100. What Schürer's works make clear is that cubature approaches give more accurate numerical integrations than QMC approaches when the integrand is continuous. It will be interesting to perform similar experiments as in [51, 52] using higher degree cubatures as such by Victoir [62] and Kuperberg [30].

Moreover, Brauchart et al. [12] discuss numerical integration errors for a certain continuous function and worst-case error for Sobolev space, by choosing several point configurations and spherical designs on $\mathbb{S}^2$, and further clarify that spherical design approaches show the best performance numerically among all other approaches under consideration. An interesting problem is to perform similar experiments as in [12] by constructing higher dimensional spherical designs.

Brauchart et al. [12] and Hirao [25] propose probabilistic generating methods of QMC design sequences using some point processes called *determinantal point processes* in probability theory. It is desirable to develop a method that is more computationally efficient than the method using *jittered samplings* proposed by Brauchart et al. [12]. To discuss QMC design sequences on integral domains (especially, "asymmetric" domain) other than cubes and spheres is still of great interest.

Isometric embeddings of Banach spaces are closely tied with the theory of cubature formula. Let p be a positive integer, consider the p-norm $\|\cdot\|_p : \mathbb{R}^d \to \mathbb{R}$ defined by

$$\|\omega\|_p = \left(\sum_{\ell=1}^d |\omega_\ell|^p \right)^{1/p}, \quad \omega = (\omega_1, \dots, \omega_d) \in \mathbb{R}^d.$$

The Euclidean space $\mathbb{R}^d$ endowed with norm $\|\cdot\|_p$ is a classical finite-dimensional Banach space, usually denoted by ℓ_p^d.

One of the important problems in the Banach space theory is to clarify the existence of an $\mathbb{R}$-linear map $\iota : \ell_2^d \to \ell_{2e}^n$ satisfying

$$\|\iota(\omega)\|_{2e} = \|\omega\|_2 \quad \text{for all } \omega \in \ell_2^d.$$

Such a map ι is called an *isometric embedding* $\ell_2^d \hookrightarrow \ell_{2e}^n$; see, e.g., [50].

A famous result in the theory of Euclidean design is the equivalence between an isometric embedding $\ell_2^d \hookrightarrow \ell_{2e}^n$ and a *Euclidean design of index $2e$* for spherical integration, i.e.,

$$\frac{1}{|\mathbb{S}^{d-1}|} \int_{\mathbb{S}^{d-1}} f(\omega)\, \rho(d\omega) = \sum_{x \in X} w(x) f(x) \quad \text{for all } f \in \mathrm{Hom}_{2e}(\mathbb{R}^d);$$

see also [32, 54] and references therein.

Example 2.11 (isometric embedding $\ell_2^3 \hookrightarrow \ell_2^4$) Here is a simple example of constructing an isometric embedding from a Euclidean design on unit sphere $\mathbb{S}^2$. Let

$$X = \{x_1, x_2, x_3, x_4\} = \left\{ \frac{1}{\sqrt{3}}(1, 1, 1), \frac{1}{\sqrt{3}}(-1, 1, 1), \frac{1}{\sqrt{3}}(1, -1, 1), \frac{1}{\sqrt{3}}(1, -1, 1) \right\}$$

be the vertices of a regular tetrahedron on unit sphere $\mathbb{S}^2$. Then X is not only a tight spherical 3-design of $\mathbb{S}^2$ but also a Euclidean design of index 2 on $\mathbb{S}^2$, i.e., it holds that

$$\frac{1}{|\mathbb{S}^2|} \int_{\mathbb{S}^2} f(\omega)\, \rho(d\omega) = \frac{1}{4} \sum_{\ell=1}^{4} f(x_\ell) \quad \text{for all } f \in \mathrm{Hom}_2(\mathbb{R}^3).$$

Then the linear map defined as

$$\iota(\omega) = \frac{\sqrt{3}}{2} (\langle \omega, x_1 \rangle, \langle \omega, x_2 \rangle, \langle \omega, x_3 \rangle, \langle \omega, x_4 \rangle) \quad \text{for all } \omega \in \mathbb{R}^3$$

is an isometric embedding $\ell_2^3 \hookrightarrow \ell_2^4$, since for $f(\omega') = \langle \omega, \omega' \rangle^2 \in \mathrm{Hom}_2(\mathbb{R}^3)$,

$$\sum_{\ell=1}^{4} w_\ell \langle \omega, x_\ell \rangle^2 = \frac{1}{|\mathbb{S}^2|} \int_{\mathbb{S}^2} \langle \omega, \omega' \rangle^2 \, \rho(d\omega') = \frac{\sqrt{3}}{2} \langle \omega, \omega \rangle^2 = \frac{\sqrt{3}}{2} \|\omega\|_2^2. \quad (2.12)$$

In particular, Shatalov [54] puts a vast amount of data of isometric embeddings together as several tables, which are called "Shatalov tables" by some authors [44, 50].

Identities of type (2.12) are called the *Hilbert identities* in the theory of quadratic and higher degree forms.

Proposition 2.3 (cf. [24, 47]) *For each $i = 1, \ldots, n$, let $x_i = (x_{i1}, \ldots, x_{id}) \in \mathbb{S}^{d-1}$ and w_i be positive reals. Let $C_{d,e}$ be a constant given by*

$$C_{d,e} = \frac{1}{|\mathbb{S}^{d-1}|} \int_{\mathbb{S}^{d-1}} \omega_1^{2e} \, \rho(d\omega) = \prod_{j=1}^{e} \frac{d + 2e - 2j}{2e + 1 - 2j}$$

with $\omega = (\omega_1, \ldots, \omega_d)$. Then the following are equivalent:

(i)   There exists an n-point Euclidean design of index $2e$ on unit sphere $\mathbb{S}^{d-1}$.
(ii) There exists a Hilbert identity as

$$C_{d,e}(\omega_1^2 + \cdots + \omega_n^2)^e = \sum_{i=1}^{n} \lambda_i (y_{i1}\omega_1 + \cdots + y_{id}\omega_d)^{2e}. \quad (2.13)$$

A particularly interesting class of Hilbert identities is that of rational identities, i.e., polynomial identities of type (2.13) with all λ_i and y_{ij} rational. For example,

$$6(x_1^2 + x_2^2 + x_3^2 + x_4^2)^2 = \sum_{1 \le i < j \le 4} (x_i + x_j)^4 + \sum_{1 \le i < j \le 4} (x_i - x_j)^4 \quad (2.14)$$

is a rational identity in four indeterminates x_1, x_2, x_3, x_4. This identity is used to prove the following type of statements in number theory (the *Waring problem* for quartic numbers): every positive integer n is expressed as a sum of at most 53 quartics. In fact, let $n = 6s + r$, $0 \le r \le 5$, and use the Lagrange four-square theorem. Then there exist integers y_{ij} such that

$$n = 6s + r = 6 \sum_{i=1}^{4} \left(\sum_{j=1}^{4} y_{ij}^2 \right)^2 + r.$$

By (2.14), each form $6 \left(\sum_{j=1}^{4} y_{ij}^2 \right)^2$ is replaced by at most 12 quartics and so n is a sum of at most 53 quartics. In general, if a rational identity of type (2.14) is explicitly constructed, then a constructive solution of the Waring problem is obtained. The Waring problem is solved by Hilbert [24] in general, but his solution relies on complicated arguments from analysis and is not constructive (see also [47, 49]).

References

1. Bannai, E.: On antipodal euclidean tight $(2e + 1)$-designs. J. Algebr. Comb. **24**(4), 391–414 (2006)
2. Bannai, E., Bannai, E.: Spherical designs and Euclidean designs. In: Recent Developments in Algebra and Related Areas (Beijing, 2007). Advance Lectures in Mathematics, vol. 8, pp. 1–37. Higher Education Press, Beijing; International Press (2009)
3. Bannai, E., Bannai, E.: A survey on spherical designs and algebraic combinatorics on spheres. Eur. J. Comb. **30**(6), 1392–1425 (2009)
4. Bannai, E., Bannai, E.: Euclidean designs and coherent configurations. In: Combinatorics and graphs. Contemporary Mathematics, vol. 531, pp. 59–93. AMS, Providence, RI (2010)
5. Bannai, E., Bannai, E., Hirao, M., Sawa, M.: Cubature formulas in numerical analysis and Euclidean tight designs. Eur. J. Comb. **31**(2), 423–441 (2010)
6. Bannai, E., Bannai, E., Hirao, M., Sawa, M.: On the existence of minimum cubature formulas for Gaussian measure on $\mathbb{R}^2$ of degree t supported by $[\frac{t}{4}] + 1$ circles. J. Algebr. Comb. **35**(1), 109–119 (2012)
7. Bannai, E., Munemasa, A., Venkov, B.: The nonexistence of certain tight spherical designs. Algebra i Analiz **16**(4), 1–23 (2004)
8. Bayer, C., Teichmann, J.: The proof of Tchakaloff's theorem. Proc. Am. Math. Soc. **134**(10), 3035–3040 (2006)
9. Bondarenko, A., Radchenko, D., Viazovska, M.: Optimal asymptotic bounds for spherical designs. Ann. Math. **178**, 443–452 (2013)
10. Bondarenko, A., Radchenko, D., Viazovska, M.: Well-separated spherical designs. Constr. Approx. **41**(1), 93–112 (2015)
11. Box, G.E.P., Behnken, D.: Some new three level designs for the study of quantitative variables. Technometrics **2**(4), 455–475 (1960)
12. Brauchart, J.S., Saff, E.B., Sloan, I.H., Womersley, R.S.: QMC designs: optimal order quasi-Monte Carlo integration schemes on the sphere. Math. Comput. **83**(290), 2821–2851 (2014)
13. Cools, R.: An encyclopaedia of cubature formulas. J. Complexity **19**(3), 445–453 (2003)
14. Cools, R., Mysovskikh, I.P., Schmid, H.J.: Cubature formulae and orthogonal polynomials. J. Comput. Appl. Math. **127**, 121–152 (2001)
15. Curto, R.E., Fialkow, L.A.: A duality proof of Tchakaloff's theorem. J. Math. Anal. Appl. **269**(2), 519–532 (2002)
16. Dao, T., De Sa, C., Ré, C.: Gaussian quadrature for kernel features. Adv. Neural. Inf. Process. Syst. **30**, 6109–6119 (2017)
17. Delsarte, P., Goethals, J.M., Seidel, J.J.: Spherical codes and designs. Geom. Dedicata **6**(3), 363–388 (1977)
18. Delsarte, P., Seidel, J.J.: Fisher type inequalities for Euclidean t-designs. Lin. Algebra Appl. **114–115**, 213–230 (1989)

19. Dick, J., Pillichshammer, F.: Digital Nets and Sequences: Discrepancy Theory and Quasi-monte Carlo Integration. Cambridge University Press, Cambridge (2010)
20. Dunkl, C.F., Xu, Y.: Orthogonal Polynomials of Several Variables. Encyclopedia of Mathematics and its Applications, vol. 155, second edn. Cambridge University Press, Cambridge (2014)
21. Engels, H.: Numerical Quadrature and Cubature. Computational Mathematics and Applications. Academic Press, Inc. [Harcourt Brace Jovanovich, Publishers], London, New York (1980)
22. Folland, G.: How to integrate a polynomial over a sphere. Amer. Math. Monthly **108**, 446–448 (2001)
23. Genz, A.: Testing multidimensional integration routines. In: Proceedings of International Conference on Tools. Methods and Languages for Scientific and Engineering Computation, pp. 81–94. North-Holland Inc, New York, NY, USA (1984)
24. Hilbert, D.: Beweis für die Darstellbarkeit der ganzen Zahlen durch eine feste Anzahl n^{ter} Potenzen (Waringsches Problem). Math. Ann. **67**(3), 281–300 (1909)
25. Hirao, M.: QMC designs and determinantal point processes. In: Monte Carlo and Quasi-monte Carlo Methods 2016, pp. 331–343. Springer, Cham (2018)
26. Hirao, M., Sawa, M.: On almost tight Euclidean designs for rotationally symmetric integrals. Jpn. J. Stat. Data Sci (To appear)
27. Hirao, M., Sawa, M.: On minimal cubature formulae of small degree for spherically symmetric integrals. SIAM J. Numer. Anal. **47**(4), 3195–3211 (2009)
28. Hirao, M., Sawa, M.: On minimal cubature formulae of odd degrees for circularly symmetric integrals. Adv. Geom. **12**(3), 483–500 (2012)
29. Krylov, V.I.: Approximate Calculation of Integrals. Dover Publications (1962)
30. Kuperberg, G.: Numerical cubature using error-correcting codes. SIAM J. Numer. Anal. **44**(3), 897–907 (2006)
31. Lyons, T., Victoir, N.: Cubature on Wiener spaces. Proc. R. Soc. Lond. A Math. Phys. Sci. **460**, 169–198 (2004)
32. Lyubich, Y.I., Vaserstein, L.N.: Isometric embeddings between classical Banach spaces, cubature formulas, and spherical designs. Geom. Dedicata **47**(3), 327–362 (1993)
33. Möller, H.M.: Polynomideale und kubaturformeln. Ph.D. thesis, University of Dortmund (1973)
34. Möller, H.M.: Kubaturformeln mit minimaler knotenzahl. Numer. Math. **35**(2), 185–200 (1976)
35. Möller, H.M.: Lower bounds for the number of nodes in cubature formulae. In: Numerische Integration (Tagung, Math. Forschungsinst., Oberwolfach, 1978). International Series of Numerical Mathematics, vol. 45, pp. 221–230. Birkhäuser, Basel-Boston, Mass (1979)
36. Mysovskih, I.P.: On the construction of cubature formulas with the smallest number of nodes (in Russian). Dokl. Akad. Nauk SSSR **178**, 1252–1254 (1968)
37. Mysovskih, I.P.: Construction of cubature formulae (in Russian). Vopr. Vychisl. i Prikl. Mat. Tashkent **32**, 85–98 (1975)
38. Mysovskih, I.P.: The approximation of multiple integrals by using interpolatory cubature formulae. In: DeVore, R.A., Scherer, K. (eds.) Quantitative Approximation. Academic Press, New York (1980)
39. Mysovskikh, I.P.: Interpolatory Cubature Formulas. Nauka, Moscow (1981). (in Russian)
40. Neumaier, A., Seidel, J.J.: Discrete measures for spherical designs, eutactic stars and lattices. Nederl. Akad. Wetensch. Indag. Math. **50**(3), 321–334 (1988)
41. Neumaier, A., Seidel, J.J.: Measures of strength $2e$ and optimal designs of degree e. Sankhyā Ser. A **54**, 299–309 (1992)
42. Niederreiter, H.: Random Number Generation and Quasi-Monte Carlo Methods, CBMS-NSF Regional Conference Series in Applied Mathematics, vol. 63. Society for Industrial and Applied Mathematics (SIAM), Philadelphia, PA (1992)
43. Noskov, M.V., Schmid, H.J.: On the number of nodes in n-dimensional cubature formulae of degree 5 for integrals over the ball. J. Comput. Appl. Math. **169**(2), 247–254 (2004)
44. Nozaki, H., Sawa, M.: Remarks on Hilbert identities, isometric embeddings, and invariant cubature. Algebra i Analiz **25**(4), 139–181 (2013)

45. Pronzato, L., Müller, W.: Design of computer experiments: space filling and beyond. Stat. Comput. **22**(3), 681–701 (2012)
46. Putinar, M.: A note on Tchakaloff's theorem. Proc. Am. Math. Soc. **125**(8), 2409–2414 (1997)
47. Reznick, B.: Sums of even powers of real linear forms. Mem. Am. Math. Soc. **96**(463) (1992)
48. Rockafellar, R.T.: Convex Analysis. Princeton University Press, Princeton (1970)
49. Sawa, M., Uchida, Y.: Discriminants of classical quasi-orthogonal polynomials with application to diophantine equations. J. Math. Soc. Jpn. (To appear)
50. Sawa, M., Xu, Y.: On positive cubature rules on the simplex and isometric embeddings. Math. Comp. **83**(287), 1251–1277 (2014)
51. Schürer, R.: Parallel high-dimensional integration: Quasi-monte carlo versus adaptive cubature rules. In: 2001 Proceedings of International Conference on Computational Science – ICCS 2001, San Francisco, CA, USA, May 28–30, Part I, pp. 1262–1271. Springer, Berlin (2001)
52. Schürer, R.: A comparison between (quasi-)monte carlo and cubature rule based methods for solving high-dimensional integration problems. Math. Comput. Simul. **62**(3–6), 509–517 (2003)
53. Seymour, P.D., Zaslavsky, T.: Averaging sets: a generalization of mean values and spherical designs. Adv. Math. **52**(3), 213–240 (1984)
54. Shatalov, O.: Isometric embeddings $l_2^m \rightarrow l_p^n$ and cubature formulas over classical fields. Ph.D. thesis, Technion-Israel Institute of Technology, Haifa, Israel (2001)
55. Silvey, S.D.: Optimal Design Monographs on Applied Probability and Statistics. Chapman & Hall, London-New York (1980)
56. Smolyak, S.A.: Quadrature and interpolation formulas for tensor products of certain classes of functions (in Russian). Dokl. Akad. Nauk SSSR **148**(5), 1042–1053 (1963)
57. Sobolev, S.L., Vaskevich, V.L.: The Theory of Cubature Formulas, Mathematics and its Applications, vol. 415. Kluwer Academic Publishers Group, Dordrecht (1997)
58. Stroud, A.H.: Approximate Calculation of Multiple Integrals. Prentice-Hall Series in Automatic Computation. Prentice-Hall Inc, Englewood Cliffs, N.J. (1971)
59. Szegő, G.: Orthogonal Polynomials. Colloquium Publications, Vol. XXIII. American Mathematical Society, Providence, R.I. (1975)
60. Tchakaloff, V.: Formules de cubatures mécaniques à coefficients non négatifs. Bull. Sci. Math. **2**(81), 123–134 (1957)
61. Verlinden, P., Cools, R.: On cubature formulae of degree $4k + 1$ attaining Möller's lower bound for integrals with circular symmetry. Numer. Math. **61**(3), 395–407 (1992)
62. Victoir, N.: Asymmetric cubature formulae with few points in high dimension for symmetric measures. SIAM J. Numer. Anal. **42**(1), 209–227 (2004)
63. Xu, Y.: Constructing cubature formulae by the method of reproducing kernel. Numer. Math. **85**(1), 155–173 (2000)
64. Xu, Y.: Lower bound for the number of nodes of cubature formulae on the unit ball. J. Complexity **19**(3), 392–402 (2003)
65. Yudin, V.A.: Lower bounds for spherical designs. Izv. Ross. Akad. Nauk Ser. Mat. **61**(3), 213–223 (1997)

Chapter 3
Optimal Euclidean Design

The main problem in optimal design theory is to find a finite set of observation points (*design*) where unknown parameters of a regression model are "well-estimated" with some statistical criterion. *D-optimality* is a popular criterion that seeks for designs minimizing the determinant of the covariance matrix. Here and hereafter, we are mainly concerned with D-optimal designs on the unit ball.

Section 3.1 starts with a quick review of polynomial regression and optimality criteria and then outlines the *Kiefer–Wolfowitz equivalence theorem* [18] that makes a beautiful connection between D- and G-optimality criteria. In Sect. 3.2, a D-optimal design on the unit ball is characterized as a weighted sum of uniform measures supported by finitely many concentric spheres (Theorem 3.3), which results in defining a certain concept called *D-optimal Euclidean design*. Section 3.3 describes ways of determining a weight set as well as a radius set of concentric spheres corresponding to D-optimal Euclidean designs. Finally, Sect. 3.4 is closed with further remarks concerning optimality criteria other than D-optimality.

While, in standard textbooks (e.g., [21]), optimal design theory is often discussed in the framework of convex analysis, the present chapter and subsequent Chap. 4 strongly emphasize connections to the theories of reproducing kernel and cubature formula.

3.1 Regression and Optimality

At first, a review of polynomial regression in statistics is given. Among many other comprehensive monographs, the interested reader may refer to, e.g., [1, 3, 6, 21].

Let Ω be a subset of $\mathbb{R}^d$ as an experimental region. Given a positive integer t, let $k = \dim \mathscr{P}_t(\Omega)$. Let $f_1, \ldots, f_k$ be a basis of the polynomial space $\mathscr{P}_t(\Omega)$ as regression functions and $F = (f_1, \ldots, f_k)^T$ be the k-dimensional column vector of these functions. With unknown real parameters $\theta_1, \ldots, \theta_k$, let $\theta = (\theta_1, \ldots, \theta_k)^T$ be the k-dimensional column vector of these parameters. Further, let Y be the response

© The Author(s), under exclusive license to Springer Nature Singapore Pte Ltd. 2019
M. Sawa et al., *Euclidean Design Theory*, JSS Research Series in Statistics,
https://doi.org/10.1007/978-981-13-8075-4_3

at a point $\omega \in \Omega$. Consider the polynomial regression model as follows:

$$Y(\omega) = F(\omega)^T \theta + \varepsilon_\omega \tag{3.1}$$

where ε_ω is a random variable with mean 0 and variance $\sigma^2 > 0$. Moreover, different observations are assumed to be independent, that is, $\mathbf{E}[\varepsilon_\omega \varepsilon_{\omega'}] = \sigma^2 \delta_{\omega,\omega'}$ for all ω, ω'.

For a set of n observations $X = \{x_1, \ldots, x_n\} \subset \Omega$, model (3.1) is represented as

$$Y(x_i) = F(x_i)^T \theta + \varepsilon_i, \quad i = 1, \ldots, n$$

where $\varepsilon_i = \varepsilon_{x_i}$ for $i = 1, \ldots, n$. These equations are also represented in matrix form as

$$\mathbf{Y} = \mathbf{X}\theta + \varepsilon$$

where

$$\mathbf{X} = (F(x_1)^T, \cdots, F(x_n)^T)^T = \begin{pmatrix} f_1(x_1) & \cdots & f_k(x_1) \\ \vdots & & \vdots \\ f_1(x_n) & \cdots & f_k(x_n) \end{pmatrix}, \quad \mathbf{Y} = (Y(x_1), \cdots, Y(x_n))^T$$

and $\varepsilon = (\varepsilon_1, \ldots, \varepsilon_n)^T$ with

$$\mathbf{E}[\varepsilon] = (0, \ldots, 0)^T, \quad \mathrm{Cov}[\varepsilon] = \sigma^2 E_n.$$

Here, E_n is the identity matrix of order n.

Particularly, $\mathbf{X}$ is called a *design matrix* for model (3.1). It is well known (e.g., [21]) that, if $\mathbf{X}^T \mathbf{X}$ is non-singular, the least squares estimator of θ in model (3.1) is given by

$$\hat{\theta} = (\mathbf{X}^T \mathbf{X})^{-1} \mathbf{X}^T \mathbf{Y},$$

which is the best linear unbiased estimator of θ by the well-known Gauss–Markov theorem. The expected value of $\hat{\theta}$ and the covariance matrix of $\hat{\theta}$ are obtained as

$$\mathbf{E}[\hat{\theta}] = \theta, \quad \mathrm{Cov}[\hat{\theta}] = \sigma^2 (\mathbf{X}^T \mathbf{X})^{-1}. \tag{3.2}$$

The matrix $\mathbf{X}^T \mathbf{X}$ is called the *information matrix* of θ for model (3.1). Note that covariance matrix $\mathrm{Cov}[\hat{\theta}]$ depends only on information matrix $\mathbf{X}^T \mathbf{X}$, i.e., depending only on the choice of X.

Example 3.1 (*Simple linear regression*) A simple linear regression is the most useful approach to predict response Y based on one variable u, i.e., model (3.1) is represented as

$$Y(u) = \theta_1 + \theta_2 u + \varepsilon_u.$$

Given a set of n observations $X = \{x_1, \ldots, x_n\}$, let

$$\mathbf{X} = \begin{pmatrix} 1 & \cdots & 1 \\ x_1 & \cdots & x_n \end{pmatrix}^T, \quad \mathbf{Y} = (Y(x_1), \cdots, Y(x_n))^T.$$

Since the information matrix is calculated by $\mathbf{X}^T\mathbf{X} = \begin{pmatrix} n & \sum_i x_i \\ \sum_i x_i & \sum_i x_i^2 \end{pmatrix}$, the least square estimator $\hat{\theta}$ of θ is calculated by

$$\hat{\theta} = (\hat{\theta}_1, \hat{\theta}_2)^T = (\mathbf{X}^T\mathbf{X})^{-1}\mathbf{X}^T\mathbf{Y}$$

$$= \frac{1}{n \sum_i x_i^2 - \sum_i (2x_i)^2} \begin{pmatrix} \sum_i x_i^2 & -\sum_i x_i \\ -\sum_i x_i & -n \end{pmatrix} \begin{pmatrix} \sum_i Y(x_i) \\ \sum_i x_i Y(x_i) \end{pmatrix}.$$

Thus

$$\hat{\theta}_1 = \frac{\sum_i x_i^2 \sum_i Y_i - \sum_i x_i \sum_i x_i Y_i}{n \sum_i x_i^2 - (\sum_i x_i)^2}, \quad \hat{\theta}_2 = \frac{n \sum_i x_i Y_i - \sum_i x_i \sum_i Y_i}{n \sum_i x_i^2 - (\sum_i x_i)^2}.$$

Now, a *design* ξ is a probability measure on Ω. Given a design ξ, a semi-definite inner product on the measure space $(\mathscr{P}_t(\Omega), \xi)$ is always defined as

$$(f, g)_\xi = \int_\Omega f(\omega) g(\omega) \, \xi(d\omega) \quad \text{for all } f, g \in \mathscr{P}_t(\Omega). \tag{3.3}$$

A design ξ is *of degree* t if a bilinear form[1] $(\cdot, \cdot)_\xi$ on $\mathscr{P}_t(\Omega)$ is positive definite, i.e., $(\cdot, \cdot)_\xi$ forms an inner product on $\mathscr{P}_t(\Omega)$. As mentioned in Sect. 1.1, the information matrix of θ using a design ξ for model (3.1), which is a generalization of information matrix $\mathbf{X}^T\mathbf{X}$, becomes a Gram matrix of $F = (f_1, \ldots, f_k)^T$.

Definition 3.1 (*Information matrix*) Given a design ξ on Ω, the $k \times k$-matrix

$$\mathbf{M}(\xi) = \int_\Omega F(\omega) F(\omega)^T \, \xi(d\omega) = \left((f_i, f_j)_\xi \right)_{i,j=1,\ldots,k}$$

is the *information matrix of* θ *using* ξ for model (3.1).

Note that, for a design ξ of degree t, the information matrix $\mathbf{M}(\xi)$ is non-singular, i.e., the inverse of $\mathbf{M}(\xi)$ is obtainable.

Example 3.2 Let X be the vertices of a regular 5-gon on $\Omega = \mathbb{S}^1$. By recalling Example 2.8, X is a spherical 4-design of $\mathbb{S}^1$. Further, let $\xi_X = \sum_{x \in X} \delta_x / 5$ be the design corresponding to X, where δ_x is the Dirac measure at point x. Then, it is easy to check that ξ_X is a design of degree 2.

---

[1] Symbols $(f, g)_\xi$ is used for $(f, g)_{\mathscr{P}_t(\Omega)}$ in the context of optimal design theory.

Set $F = (\phi_{0,1}, \phi_{1,1}, \phi_{1,2}, \phi_{2,1}, \phi_{2,2})^T$, where $\phi_{\ell,i}$ is an orthonormal basis of $\text{Harm}_\ell(\mathbb{R}^2)$ defined in Theorem 1.5. It is easy to check that the information matrix $\mathbf{M}(\xi_X)$ of θ for model (3.1) is reduced to the identity matrix E_5. Thus $\mathbf{M}(\xi_X)$ is non-singular.

Remark 3.1 For n distinct points $x_1, \ldots, x_n$ in Ω, let $\xi_n = \sum_{i=1}^n p_i \delta_{x_i}$ be a convex combination of Dirac measures δ_{x_i}. This ξ_n is called a *continuous* or an *approximate design*, as introduced by Kiefer [14]. Then, the covariance matrix of $\hat{\theta}$ and the variance of $F(\omega)^T \hat{\theta}$ at point $\omega \in \Omega$ for model (3.1) are calculated as follows:

$$\text{Cov}[\hat{\theta}] = \frac{\sigma^2}{n} \mathbf{M}^{-1}(\xi_n), \tag{3.4}$$

$$\text{Var}[F(\omega)^T \hat{\theta}] = \frac{\sigma^2}{n} d(\omega, \xi_n),$$

where $d(\cdot, \xi)$ is the corresponding normalized variance function defined by

$$d(\omega, \xi) = F(\omega)^T \mathbf{M}^{-1}(\xi) F(\omega), \quad \omega \in \Omega. \tag{3.5}$$

To explain the relationship between experimental designs and Euclidean designs given in Sect. 2.4, the two important concepts, *rotatability* and *invariance*, are introduced. Let $\mathcal{O}(d)$ be the orthogonal group of $\mathbb{R}^d$. Further let $\xi_{Q^{-1}}$ be the induced measure with respect to $Q \in \mathcal{O}(d)$, i.e., $\xi_{Q^{-1}}(A) = \xi(QA)$ for all measurable set A of Ω.

Definition 3.2 (*Rotatable design*)

(i) A design ξ is said to be *rotatable* if $(\cdot, \cdot)_\xi = (\cdot, \cdot)_{\xi_{Q^{-1}}}$ for all $Q \in \mathcal{O}(d)$.
(ii) A design ξ is said to be $\mathcal{O}(d)$-*invariant* (or *invariant* for simplicity) if $\xi = \xi_{Q^{-1}}$ for all $Q \in \mathcal{O}(d)$.

Given a design ξ, there exists an invariant design $\bar{\xi}$ as

$$\bar{\xi} = \int_{Q \in \mathcal{O}(d)} \xi_{Q^{-1}} d\mu$$

where μ is the normalized Haar measure on $\mathcal{O}(d)$. Note that invariant designs are also rotatable designs.

Remark 3.2 Rotatability is often defined in terms of $d(\cdot, \xi)$, namely, a design ξ is said to be *rotatable* if

$$d(\cdot, \xi) = d(Q\cdot, \xi) \quad \text{for all } Q \in \mathcal{O}(d).$$

Example 3.3 Recall Example 3.2. By using polar coordinates $(\omega_1, \omega_2) = (\cos\theta, \sin\theta)$ in $\mathbb{S}^1$, an orthonormal basis $\{\phi_{\ell,i}\}$ of $\text{Harm}_\ell(\mathbb{R}^2)$ is explicitly given by

$\phi_{0,1} \equiv 1, \quad \phi_{\ell,1}(\omega) = \phi_{l,1}(\theta) = \sqrt{2}\cos(\ell\theta), \quad \phi_{\ell,2}(\omega) = \phi_{l,2}(\theta) = \sqrt{2}\cos(\ell\theta), \quad \ell = 1, 2, \ldots.$

It is easy to check that

$$\begin{aligned} d(\omega, \xi_X) &= F(\theta)^T F(\theta) \\ &= 1 + (\sqrt{2}\cos\theta)^2 + (\sqrt{2}\sin\theta)^2 + (\sqrt{2}\cos(2\theta))^2 + (\sqrt{2}\sin(2\theta))^2 \\ &= 5 \quad \text{for all } \omega \in \mathbb{S}^{d-1}, \end{aligned}$$

which implies that ξ_X is a rotatable design (of degree 2). In general, a Euclidean $2e$-design is a rotatable design of degree e, and vice versa; see [19] and Proposition 3.1 in Sect. 3.2.

The present goal on optimal designs is to find a design ξ for model (3.1), which decreases covariance matrix $\mathrm{Cov}[\hat{\theta}]$ in some optimality criteria.

Many optimality criteria for design, which are explicitly or implicitly written in terms of eigenvalues of the information matrix, have been proposed; see e.g., Atkinson et al. [1], Kiefer [17], Pukelsheim [21] and Silvey [25]. The following are typical examples:

- The D-optimality criterion seeks for a design ξ such that

$$\xi \text{ minimizes } \det \mathbf{M}^{-1}(\xi) \text{ (or maximizes } \det \mathbf{M}(\xi)).$$

For D-optimal designs, the least square estimator $\hat{\theta}$ of θ in model (3.1) has the smallest determinant of the covariance matrix. This is equivalent to minimizing the volume of confidence ellipsoids for the estimator $\hat{\theta}$.

- The A-optimality criterion seeks for a design ξ such that

$$\xi \text{ minimizes } \mathrm{tr}\, \mathbf{M}^{-1}(\xi).$$

For A-optimal designs, the sum or average of variance of the least square estimator $\hat{\theta}$ of θ in model (3.1) is as small as possible.

- The G-optimality criterion seeks for a design ξ such that

$$\xi \text{ minimizes } \max_{\omega} d(\omega, \xi);$$

see also (3.5). For G-optimal designs, the largest prediction variance overall points in Ω is as small as possible.

- The E-optimality criterion seeks for a design ξ such that

$$\xi \text{ minimizes } \max_{i} \lambda_i^{-1}(\xi)$$

where $\lambda_1(\xi), \ldots, \lambda_k(\xi)$ are the eigenvalues of $\mathbf{M}(\xi)$. This criterion is known (e.g., [1, 25]) as a special case of G-optimality criterion. For E-optimal designs, the largest variance of $\omega^T\hat{\theta}$ over all ω in $\mathbb{S}^{k-1}$ is as small as possible.

Among these criteria, the next section particularly focuses on the most popular *D*-optimality criterion.

In the remaining part of this section, a relationship between *D*- and *G*-optimality criteria is discussed. Kiefer and Wolfowitz [18] (see also Karlin and Studden [13]) show the equivalence of *D*- and *G*-optimality criteria, which plays a key role to classify *D*-optimal designs on $\mathbb{B}^d$ as in Theorem 3.3.

Theorem 3.1 (Kiefer–Wolfowitz equivalence theorem) *Let* $\mathbf{M}(\xi)$ *be the information matrix of* θ *using* ξ *for model (3.1), and let* $d(\cdot, \xi)$ *be the corresponding normalized variance function for* $\mathbf{M}(\xi)$. *Then, the following three statements are equivalent:*

(i) ξ *maximizes* $\det \mathbf{M}(\xi)$.
(ii) ξ *minimizes* $\max_{\omega} d(\omega, \xi)$.
(iii) ξ *satisfies that*

$$\max_{\omega} d(\omega, \xi) = \dim \mathscr{P}_t(\Omega).$$

Moreover, the set Ξ *of all* ξ *satisfying these conditions is convex, and* $\mathbf{M}(\xi)$ *does not depend on the choice of* ξ *in* Ξ.

Corollary 3.1 *Given a positive integer t, let* ξ *be a D-optimal design of degree t. Then, it holds that*

$$d(\omega, \xi) = \dim \mathscr{P}_t(\Omega) \quad \text{for all } \omega \in \text{supp}(\xi).$$

Proof Let K be the kernel function (reproducing kernel) of space $\mathscr{P}_t(\Omega)$ equipped with inner product $\int_{\Omega} \cdot \, d\xi$. Then, by (1.3), function $d(\omega, \xi)$ can be expressed in terms of kernel K as follows:

$$d(\omega, \xi) = K(\omega, \omega) \quad \text{for all } \omega \in \Omega.$$

It follows from Proposition 1.4 that

$$\int_{\Omega} d(\omega, \xi) \, \xi(d\omega) = \int_{\Omega} K(\omega, \omega) \, \xi(d\omega) = \dim \mathscr{P}_t(\Omega).$$

Therefore,

$$\max_{\omega} d(\omega, \xi) \geq \dim \mathscr{P}_t(\Omega).$$

The result then follows from Theorem 3.1. □

Note that other criteria and the Kiefer general equivalence theorem [16] are also discussed in Sect. 3.4.

3.2 Optimal Euclidean Design and Characterization Theorems

Throughout this section, let us consider d-dimensional unit ball $\mathbb{B}^d = \{\omega \in \mathbb{R}^d \mid \|\omega\| \leq 1\}$ as an $\mathcal{O}(d)$-invariant experimental region. As usual, the standard eth degree polynomial regression model is considered. Namely, functions $f_1, \ldots, f_k$ in mode (3.1) is a basis of $\mathscr{P}_e(\mathbb{B}^d)$ with $k = \dim \mathscr{P}_e(\mathbb{B}^d)$.

The following is well known in the theory of optimal experimental design.

Theorem 3.2 (cf. [15, 19]) *Assume that there exists a D-optimal design (for model (3.1)) of degree e on $\mathbb{B}^d$. Then, the following (i) and (ii) hold:*

(i) *All D-optimal designs of degree e have the same inner product on $\mathscr{P}_e(\mathbb{B}^d)$, i.e., have the same information matrix $\mathbf{M}(\xi)$.*
(ii) *The set of all D-optimal designs of degree e is a closed convex set consisting of rotatable designs, and containing an invariant design.*

Since any invariant design ξ on $\mathbb{B}^d$ can be separated as

$$\xi(B) = \int_0^1 \rho(r^{-1}(B \cap \mathbb{S}_r^{d-1}))\, \tau(dr) \tag{3.6}$$

where τ is a probability measure on interval $[0, 1]$ and B is any measurable set of $\mathbb{S}_r^{d-1}$.

The following proposition implies that any invariant design on the unit ball can be represented as a probability measure supported by finitely many concentric spheres centered at the origin.

Proposition 3.1 *Let ξ be an invariant design of degree e on $\mathbb{B}^d$. Assume that there exists a finite weighted pair $(\{r_i\}_i, \{W_i\}_i)$ satisfying*

$$\int_{\mathbb{B}^d} f(\omega)\, \xi(d\omega) = \sum_{i=1}^p \frac{W_i}{|\mathbb{S}_{r_i}^{d-1}|} \int_{\mathbb{S}_{r_i}^{d-1}} f(\omega)\, \rho_{r_i}(d\omega) \quad \text{for all } f \in \mathscr{P}_{2e}(\mathbb{R}^d).$$

Then it holds that $p \geq \lfloor e/2 \rfloor + 1$.

Sketch of proof. A key point of this proof is to consider quadratures corresponding to the radial components of ξ.

By recalling the argument used in the proof of Theorem 2.8 and (3.6), it suffices to find a pair $(\{r_i\}_i, \{W_i\}_i)$ satisfying

$$\int_0^1 r^{2\ell}\, \tau(dr) = \sum_{i=1}^p W_i r_i^{2\ell}, \quad \ell = 0, 1, \ldots, e.$$

Namely, $(\{r_i^2\}_i, \{W_i\}_i)$ forms a quadrature formula of degree e. The required result follows from (2.1) or (2.2). $\qquad\square$

The following forms a key result for further arguments developed in Chap. 4.

Theorem 3.3 (Kiefer characterization theorem, [15]) *Let ξ be a D-optimal invariant design of degree e on $\mathbb{B}^d$. Then there uniquely exist $\mathscr{R} = \{r_i\}_{i=1}^{\lfloor e/2\rfloor+1}$ and $\mathscr{W} = \{W_i\}_{i=1}^{\lfloor e/2\rfloor+1}$, such that $\sum_{i=1}^{\lfloor e/2\rfloor+1} W_i = 1$ and*

$$
\int_{\mathbb{B}^d} f(\omega)\,\xi(d\omega) = \sum_{i=1}^{\lfloor e/2\rfloor+1} \frac{W_i}{|\mathbb{S}_{r_i}^{d-1}|} \int_{\mathbb{S}_{r_i}^{d-1}} f(\omega)\,\rho_{r_i}(d\omega) \quad \text{for all } f \in \mathscr{P}_{2e}(\mathbb{R}^d)
$$

with $r_1 = 1 > r_2 > \cdots r_{\lfloor e/2\rfloor+1} \geq 0$, and $r_{\lfloor e/2\rfloor+1} = 0$ when e is even.

Sketch of proof. The proof below is based on Proposition 3.1 and Corollary 3.1, which is a blended version of two proofs, one by Farrell et al. [9], and one by Neumaier and Seidel [19].[2]

Assume that $e = 3$ and F is an orthonormal basis of $\mathscr{P}_3(\mathbb{B}^d)$ for simplicity. By Theorem 3.1 and Corollary 3.1, it is seen that a design ξ^* is D-optimal if and only if

$$
d(\omega, \xi^*) = F(\omega)^T \mathbf{M}^{-1}(\xi^*) F(\omega) \leq k = \binom{d+3}{3} \quad \text{for all } \omega \in \mathbb{B}^d
$$

with the equality for points on the supports of ξ^*. Since $F(\omega)$ gives an orthonormal basis, for any $Q \in \mathscr{O}(d)$ there exists an orthogonal matrix U_Q such that $F(Q\omega) = F(\omega)U_Q$. Then it holds that

$$
\begin{aligned}
d(Q\omega, \xi^*) &= F(Q\omega)\mathbf{M}^{-1}(\xi^*)F(Q\omega)^T \\
&= F(\omega)U_Q\mathbf{M}^{-1}(\xi^*)U_Q'F(\omega)^T \\
&= F(\omega)\mathbf{M}^{-1}(\xi_{Q^{-1}}^*)F(\omega)^T \\
&= F(\omega)\mathbf{M}^{-1}(\xi^*)F(\omega)^T \\
&= d(\omega, \xi^*).
\end{aligned}
$$

It thus turns out (cf. [7, 9]) that

$$
d(\omega, \xi^*) = \alpha + \beta\|\omega\|^2 + \gamma\|\omega\|^4 + \delta\|\omega\|^6 \tag{3.7}
$$

where $\alpha, \beta, \gamma, \delta$ do not depend on the choice of ω. Since $\mathbf{M}(\xi^*)$ is non-singular, δ must be positive and $d(\omega, \xi^*)$ is a cubic polynomial of $\|\omega\|^2$. There are at most two maximum points on $0 \leq \|\omega\| \leq 1$, and hence by Proposition 3.1, the number of spheres must be two. Finally, it is easy to see that one of the spheres has radius one by (3.7).   □

---

[2]A different proof can be found in Karlin and Studden [13] where techniques concerning Hankel matrices and convexity are employed.

Remark 3.3 Replacing an invariant design of type $\sum_{i=1}^{\lfloor e/2 \rfloor + 1} W_i \rho_{r_i} / |\mathbb{S}_{r_i}^{d-1}|$, *D*-optimal or not, by a weighted point set is just the concept of Euclidean design. In Draper and Pukelsheim [8], a weighted uniform measure $\sum_{i=1}^{\lfloor e/2 \rfloor + 1} W_i \rho_{r_i} / |\mathbb{S}_{r_i}^{d-1}|$ is called a "boundary nucleus design".

By recalling Corollary 2.1, a cubature formula for an integral with respect to a *D*-optimal invariant measure ξ^* is a Euclidean design. Let us now define the *D*-optimality for Euclidean 2*e*-designs, first introduced by Bannai and Bannai [2].

Definition 3.3 (*D-optimal Euclidean design*) A pair (X, w) is called a *D-optimal Euclidean 2e-design on* $\mathbb{B}^d$, if it is a Euclidean 2*e*-design supported by $\lfloor e/2 \rfloor + 1$ concentric spheres and there exists a *D*-optimal invariant design ξ^* of degree *e* on $\mathbb{B}^d$ such that

$$\int_{\mathbb{B}^d} f(\omega)\, \xi^*(d\omega) = \sum_{i=1}^{\lfloor e/2 \rfloor + 1} \frac{W_i}{|\mathbb{S}_{r_i}^{d-1}|} \int_{\mathbb{S}_{r_i}^{d-1}} f(\omega)\, \rho_{r_i}(d\omega) \quad \text{for all } f \in \mathscr{P}_{2e}(\mathbb{R}^d).$$

Example 3.4 Two concrete examples of *D*-optimal Euclidean 4- and 6-designs on $\mathbb{B}^2$ are, respectively, given as

$$\mathscr{Q}[f] = \frac{1}{6} f(0, 0) + \frac{1}{6} \sum_{\ell=0}^{4} f\left(\cos\frac{2\ell\pi}{5}, \sin\frac{2\ell\pi}{5}\right), \tag{3.8}$$

$$\mathscr{Q}[f] = \frac{1 - W_2}{5} \sum_{\ell=0}^{4} f\left(\cos\frac{2\ell\pi}{5}, \sin\frac{2\ell\pi}{5}\right) + \frac{W_2}{5} \sum_{\ell=0}^{4} f\left(r_2 \cos\frac{(2\ell + 1)\pi}{5}, r_2 \sin\frac{(2\ell + 1)\pi}{5}\right)$$

where W_2 and r_2 are some optimal values; see Fig. 3.1. Details on how to calculate these optimal weight W_2 and radius r_2 are described in Sect. 2.3. Here, a *D*-optimal Euclidean 4-design is focused. Since a set $\{1, \|\omega\|^2, \phi_{1,1}, \phi_{1,2}, \phi_{2,1}, \phi_{2,2}\}$ forms an orthonormal basis of $\mathscr{P}_2(\mathbb{B}^2)$ (see Theorem 1.5 for notation $\phi_{\ell,i}$), we can let $F = (1, \|\omega\|^2, \phi_{1,1}, \phi_{1,2}, \phi_{2,1}, \phi_{2,2})^T$; see also (3.11) for more details. Further, let ξ_X be the corresponding measure of the *D*-optimal Euclidean 4-design defined in (3.8). Then, it is easy to check that the information matrix using ξ_X is calculated as

$$\mathbf{M}(\xi_X) = \begin{pmatrix} 1 & 5/6 & 0 & 0 & 0 & 0 \\ 5/6 & 5/6 & 0 & 0 & 0 & 0 \\ 0 & 0 & 5/6 & 0 & 0 & 0 \\ 0 & 0 & 0 & 5/6 & 0 & 0 \\ 0 & 0 & 0 & 0 & 5/6 & 0 \\ 0 & 0 & 0 & 0 & 0 & 5/6 \end{pmatrix}.$$

Thus, the normalized variance function $d(\cdot, \xi_X)$ is represented as

$$d(\omega, \xi_X) = 6 - \frac{48}{5}\|\omega\|^2 + \frac{48}{5}\|\omega\|^4 \le 6,$$

where the equality holds if $\omega \in \{(0, 0)\} \cup \mathbb{S}^1$; see also Fig. 3.2.

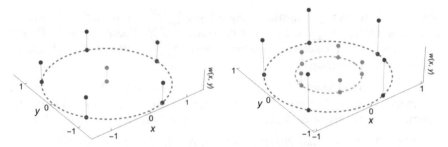

Fig. 3.1 *D*-optimal Euclidean 4-design (left) and 6-design (right) of $\mathbb{B}^2$

Fig. 3.2 Normalized
variance function $d(\omega, \xi_X)$
with respect to *D*-optimal
Euclidean 4-design

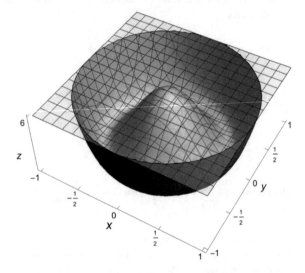

In Chap. 4, *D*-optimal Euclidean designs are constructed by using *corner vector methods*.

In the remaining part of this section, bounds for *D*-optimal Euclidean 2*e*-designs are briefly reviewed.

Proposition 3.2 (cf. [9]) *Given positive integers d and e, let*

$$N_{d,e} := \min \left\{ |X| \mid X \text{ is the point set of a D-optimal Euclidean 2e-design on } \mathbb{B}^d \right\}.$$

Then

$$\binom{d+e}{e} \leq N_{d,e} \leq \binom{d+2e}{2e}. \tag{3.9}$$

By Theorem 2.1, there exists a *D*-optimal Euclidean 2*e*-design (X, w) on $\mathbb{B}^d$ supported by $\lfloor e/2 \rfloor + 1$ concentric spheres such that

$$|X| \le \binom{d+2e}{2e} (= \dim \mathscr{P}_{2e}(\mathbb{B}^d)) \tag{3.10}$$

which implies the right-hand side of (3.9). On the other hand, the Fisher-type bound (2.1) tells us

$$|X| \ge \binom{d+e}{e} (= \dim \mathscr{P}_e(\mathbb{B}^d))$$

showing the left-hand side of (3.9). This can be interpreted in a statistical manner: Since model (3.1) has $N \left(= \binom{d+e}{e}\right)$ unknown parameters θ, in order to estimate these parameters, we need at least N observation locations.

These results are classical both in numerical analysis and design of experiments, e.g., in optimal design theory, which seem not to have been fully recognized in combinatorics, until relatively recently.

3.3 Realization of the Kiefer Characterization Theorem

As seen in the previous section, the Kiefer theorem (Theorem 3.3) characterizes a D-optimal invariant design by a weighted summation of uniform measures supported by some concentric spheres. As a realization of this fact, the present section discusses how to calculate radii and weights of D-optimal invariant designs. These information will be employed when developing construction methods of D-optimal Euclidean designs on unit ball $\mathbb{B}^d$ in the next chapter.

The following proposition plays a key role to classify D-optimal designs.

Proposition 3.3 *A D-optimal design for the eth degree polynomial regression model (3.1) on $\mathbb{B}^d$ does not depend on the choice of a basis of $\mathscr{P}_e(\mathbb{B}^d)$.*

Proof Let F and G be two k-dimensional column vectors of bases of $\mathscr{P}_e(\mathbb{B}^d)$ with $k = \dim \mathscr{P}_e(\mathbb{B}^d)$, respectively. Further, let A be a matrix of the transformation defined by $F = AG$. Then the information matrix of θ using ξ for model (3.1) is

$$\mathbf{M}(\xi) = \int_{\mathbb{B}^d} F(\omega)F(\omega)^T \, \xi(d\omega) = A \int_{\mathbb{B}^d} G(\omega)G(\omega)^T \, \xi(d\omega) A^T = A\widetilde{\mathbf{M}}(\xi)A^T$$

where $\widetilde{\mathbf{M}}(\xi) = \int_{\mathbb{B}^d} G(\omega)G(\omega)^T \, \xi(d\omega)$. Thus it holds that $\det \mathbf{M}(\xi) = (\det A)^2 \det \widetilde{\mathbf{M}}(\xi)$. If a design ξ maximizes $\det \mathbf{M}(\xi)$, then ξ also maximizes $\widetilde{\mathbf{M}}(\xi)$ since $\det F$ is independent of ξ. Thus, a D-optimum design is determined regardless of the choice of bases of $\mathscr{P}_e(\mathbb{B}^d)$. $\qquad\square$

Following Bannai and Bannai [2] and Neumaier and Seidel [19], a multivariate "Zernike-type" polynomial basis of $\mathscr{P}_e(\mathbb{B}^d)$ is adopted hereinafter to specify D-optimal designs. By recalling $\phi_{\ell,i}$ and h_ℓ^d defined in Sect. 1.2, a Zernike-type polynomial basis of $\mathscr{P}_e(\mathbb{B}^d)$ is given as

$$\mathcal{D}_e(\mathbb{B}^d) = \left\{ \|\omega\|^{2j}\phi_{\ell,i} \mid 0 \le \ell \le e,\ 0 \le j \le \left\lfloor \frac{e-\ell}{2} \right\rfloor,\ 1 \le i \le h_\ell^d \right\}. \quad (3.11)$$

As will be described along the lines given in Bannai and Bannai [2] (see also Sawa and Hirao [24]), an advantage of this basis is the simplicity for calculating information matrices.

From Theorem 3.3, it is seen that there exists a D-optimal invariant design ξ^* of degree e supported by $\lfloor e/2 \rfloor + 1$ concentric spheres $S_{\lfloor e/2 \rfloor+1}$, which has the form

$$\int_{\mathbb{B}^d} f(\omega)\,\xi^*(d\omega) = \sum_{i=1}^{\lfloor e/2 \rfloor+1} \frac{W_i}{|\mathbb{S}^{d-1}|} \int_{\mathbb{S}^{d-1}} f(r_i\omega)\,\rho(d\omega) \quad (3.12)$$

where $\sum_i W_i \equiv 1$, $r_1 = 1$ and $r_{\lfloor e/2 \rfloor+1} = 0$ when e is even.

The entries of information matrix $\mathbf{M}(\xi^*)$ are given as follows.

(i) If $0 \in S_{\lfloor e/2 \rfloor+1}$, then for $\|\omega\|^{2j_1}\phi_{\ell_1,k_1},\ \|\omega\|^{2j_2}\phi_{\ell_2,k_2} \in \mathcal{D}_e(\mathbb{B}^d)$

$$(\|\omega\|^{2j_1}\phi_{\ell_1,k_1},\ \|\omega\|^{2j_2}\phi_{\ell_2,k_2})_{\xi^*}$$

$$= \begin{cases} 1 & \text{if } j_1 = j_2 = \ell_1 = \ell_2 = 0, \\ \delta_{k_1,k_2}\delta_{\ell_1,\ell_2} \sum_{i=1}^{p} r_i^{2(j_1+j_2+\ell_1)} W_i & \text{if } j_1 + j_2 + \ell_1 + \ell_2 \ge 1. \end{cases}$$

(ii) If $0 \notin S_{\lfloor e/2 \rfloor+1}$, then for $\|\omega\|^{2j_1}\phi_{\ell_1,k_1},\ \|\omega\|^{2j_2}\phi_{\ell_2,k_2} \in \mathcal{D}_e(\mathbb{B}^d)$

$$(\|\omega\|^{2j_1}\phi_{\ell_1,k_1},\ \|\omega\|^{2j_2}\phi_{\ell_2,k_2})_{\xi^*} = \delta_{k_1,k_2}\delta_{\ell_1,\ell_2} \sum_{i=1}^{\lfloor e/2 \rfloor+1} r_i^{2(j_1+j_2+\ell_1)} W_i.$$

Let $\mathbf{M}_{\ell,k}(\xi^*)$ be the matrix with (j_1, j_2)-entry defined by

$$(\|\omega\|^{2j_1}\phi_{\ell,k},\ \|\omega\|^{2j_2}\phi_{\ell,k})_{\xi^*}$$

for $\|\omega\|^{2j_1}\phi_{\ell,k},\ \|\omega\|^{2j_2}\phi_{\ell,k} \in \mathcal{D}_e(\mathbb{B}^d)$. Since it holds that for each ℓ

$$\mathbf{M}_{\ell,k_1}(\xi^*) = \mathbf{M}_{\ell,k_2}(\xi^*) \quad \text{for every } 1 \le k_1, k_2 \le h_\ell^d,$$

each $\mathbf{M}_{\ell,k}(\xi^*)$ is denoted by $\mathbf{M}_\ell(\xi^*)$. Then it can be easily checked that the determinant of $\mathbf{M}(\xi^*)$ is represented by products of determinants of $\mathbf{M}_\ell(\xi^*)$.

Lemma 3.1 *Let ξ^* be a D-optimal invariant design defined in (3.12). Then it holds that*

$$\det \mathbf{M}(\xi^*) = \prod_{i=0}^{e} \left(\det \mathbf{M}_\ell(\xi^*) \right)^{h_\ell^d}. \quad (3.13)$$

Example 3.5 (cf. [2, 24]) Let us first compute the optimal weights W_i of a D-optimal invariant design of degree 2. It holds that

$$\mathbf{M}_0(\xi^*) = \begin{pmatrix} 1 & W_1 \\ W_1 & W_1 \end{pmatrix}, \quad \mathbf{M}_1(\xi^*) = \mathbf{M}_2(\xi^*) = (W_1).$$

Equation (3.13) implies that

$$\det \mathbf{M}(\xi^*) = (1 - W_1) W_1^{d(d+3)/2}.$$

Thus it follows that

$$W_1 = \frac{d(d+3)}{(d+1)(d+2)}, \quad W_2 = \frac{2}{(d+1)(d+2)}. \tag{3.14}$$

Next, let us compute the optimal values of W_i and r_i of a D-optimal invariant design of degree 3. Since $0 \notin S_2$, it holds that

$$\mathbf{M}_0(\xi^*) = \begin{pmatrix} 1 & W_1 + r_2^2 W_2 \\ W_1 + r_2^2 W_2 & W_1 + r_2^4 W_2 \end{pmatrix}, \quad \mathbf{M}_1(\xi^*) = \begin{pmatrix} W_1 + r_2^2 W_2 & W_1 + r_2^4 W_2 \\ W_1 + r_2^4 W_2 & W_1 + r_2^6 W_2 \end{pmatrix},$$
$$\mathbf{M}_2(\xi^*) = (W_1 + r_2^4 W_2), \quad \mathbf{M}_3(\xi^*) = (W_1 + r_2^6 W_2).$$

Equation (3.13) implies that

$$\det \mathbf{M}(\xi^*) = \prod_{\ell=0}^{3} (\det \mathbf{M}_\ell(\xi^*))^{h_\ell^d}.$$

Note that $W_1 = 1 - W_2$. In order to obtain the optimal values of W_2 and r_2, it suffices to solve the following equations:

$$
\begin{cases}
\dfrac{d}{r_2^2} - \dfrac{2(d+1)}{1-r_2^2} + \dfrac{2h_2^d W_2 r_2^2}{1 - W_2 + W_2 r_2^4} + \dfrac{3h_3^d W_2 r_2^4}{1 - W_2 + W_2 r_2^6} = 0, \\[4mm]
\dfrac{(d+1)(1-2W_2)}{W_2(1-W_2)} - \dfrac{h_2^d(1-r_2^4)}{1 - W_2 + W_2 r_2^4} - \dfrac{h_3^d(1-r_2^6)}{1 - W_2 + W_2 r_2^2} = 0;
\end{cases}
\tag{3.15}
$$

see also [11, 19]. Solving them numerically generates optimal values of W_2 and r_2^2.

Now we list up in Table 3.1 all optimal values of W_2 and r_2^2 for D-optimal Euclidean 6-designs in low dimensions by use of computer software Mathematica. Of course, depending on practical situations, the desired estimation accuracy of these values will vary. However, all values of W_2 and r_2^2 listed up in Table 3.1 are rounded to no more than six significant figures. Note that throughout this book, optimal values are rounded to no more than six significant figures for convenience.

Chapter 4 will present several examples of D-optimal Euclidean 6-designs by taking these optimal values of W_2 and r_2^2; see subsequent Sects. 4.3.2, 4.3.3 and 4.5.

Table 3.1 Characteristics of D-optimal Euclidean 6-designs

d	W_2	r_2^2
3	0.208007	0.296255
4	0.149692	0.313408
5	0.112748	0.324217
6	0.087914	0.331571
7	0.070438	0.336856
8	0.0576841	0.340816
9	0.0480964	0.343882
10	0.040710	0.346316
$\vdots$	$\vdots$	$\vdots$
50	0.002297	0.362628

Remark 3.4 In calculating the optimum values, a "Zernike-type" polynomial regression model is considered here. The original Zernike polynomial regression model is defined as follows: With unknown real parameters $\{\theta_{i,j,\ell}\}$, define the model

$$Y(\omega) = \sum_{\ell=0}^{t} \sum_{j=0}^{\lfloor \frac{t-\ell}{2} \rfloor} \sum_{i=1}^{h_\ell^d} \theta_{i,j,\ell} R_{\ell,j}(\|\omega\|^2)\phi_{\ell,i}(\omega) + \varepsilon_\omega, \quad \omega \in \mathbb{B}^d,$$

where $R_{\ell,j}$ is a polynomial of $\|\omega\|^2$ of total degree $2j$ such that

$$\frac{1}{\int_0^1 r^{d-1} \, dr} \int_0^1 R_{\ell,j}(r^2) R_{\ell,j'}(r^2) r^{2\ell+d-1} \, dr = \delta_{j,j'}.$$

Optimal designs for a multivariate Zernike polynomial regression model are discussed in, e.g., Dette et al. [7] and Hirao et al. [11]. This model is useful in several areas, including optics, circular or spherical data analysis, and so on; see also Ramos-López et al. [22].

3.4   Further Remarks and Open Questions

While in the present chapter the D-optimality criterion is mainly discussed, there are many other criteria to be considered. Among them, it is theoretically important to consider the Kiefer ϕ-optimality criteria (cf. [16, 17]): Let Ξ be a class of probability measures on Ω. Further let $\mathcal{M} = \{\mathbf{M}(\xi) \mid \eta \in \Xi\}$ and ϕ be a real-valued concave function on $\mathcal{M}$. A design ξ maximizing $\phi(\mathbf{M}(\xi))$ among all ξ is called a *ϕ-optimal design*.

For example, let $\phi = \phi_q$ be a real function on $\mathcal{M}$ defined by

$$\phi_q(\mathbf{M}(\xi)) = \left(k^{-1}\mathrm{tr}(\mathbf{M}^{-q}(\xi))\right)^{1/q} = \left(k^{-1}\sum_{i=1}^{k}\lambda_i^{-q}(\xi)\right)^{1/q}, \quad q \neq 0, \ -1 \leq q < \infty$$

and

$$\phi_0(\mathbf{M}(\xi)) = \lim_{q\to+0}\phi_q(\mathbf{M}(\xi)) = \left(\det\mathbf{M}^{-1}(\xi)\right)^{1/k} = \left(\prod_{i=1}^{k}\lambda_k(\xi)\right)^{-1/k},$$

$$\phi_\infty(\mathbf{M}(\xi)) = \lim_{q\to+\infty}\phi_q(\mathbf{M}(\xi)) = \max_i \lambda_i^{-1}(\xi)$$

where $\lambda_1(\xi), \ldots, \lambda_k(\xi)$ are eigenvalues of $\mathbf{M}(\xi)$. The ϕ_0-, ϕ_1-, and ϕ_∞-optimality criteria are familiar with D-, A- and E-optimality criteria; see also Sect. 3.1.

Kiefer [16] shows the following general equivalence theorem as a full generalization of Theorem 3.1 (see also Silvey [25]): Given $M_1, M_2 \in \mathcal{M}$, the Fréchet derivative (e.g., [23]) is defined as

$$\mathscr{F}_\phi(M_1, M_2) = \lim_{\varepsilon\to+0}\frac{1}{\varepsilon}\left(\phi\left((1-\varepsilon)M_1 + \varepsilon M_2\right) - \phi(M_1)\right).$$

Theorem 3.4 (cf. [16, 25]) *Assume that ϕ is concave on $\mathcal{M}$. Then ξ^* is ϕ-optimal if and only if*

$$\mathscr{F}_\phi(\mathbf{M}(\xi^*), \mathbf{M}(\xi)) \leq 0 \quad \text{for all } \xi \in \varXi.$$

Moreover, when ϕ is concave on $\mathcal{M}$ and differentiable at $\mathbf{M}(\xi')$,

$$\mathscr{F}_\phi(\mathbf{M}(\xi^*), F(\omega)F(\omega)^T) \leq 0 \quad \text{for all } \omega \in \varOmega.$$

Example 3.6 ([25]) Let $\mathcal{M}$ be a set of information matrices of θ for model (3.1) and $\phi(M) = \log\det M$ be a real function on $M \in \mathcal{M}$. By using determinant approximations, it is easy to check that the Gâteaux derivative of $\log\det M$ at M_1 in the direction of M_2 (e.g., [23]) is

$$\mathscr{G}_\phi(M_1, M_2) = \lim_{\varepsilon\to+0}\frac{\log\det(M_1 + \varepsilon M_2) - \log\det M_1}{\varepsilon} = \mathrm{tr}\, M_2 M_1^{-1}.$$

Thus it holds that

$$\mathscr{F}_\phi(M_1, M_2) = \mathscr{G}_\phi(M_1, M_2 - M_1) = \mathrm{tr}\, M_2 M_1^{-1} - k.$$

Thus, by using Theorem 3.4, ξ is D-optimal if and only if

$$\mathscr{F}_\phi(\mathbf{M}(\xi), F(\omega)F(\omega)^T) = F(\omega)^T\mathbf{M}^{-1}(\xi)F(\omega) - k = d(\omega, \xi) - k \leq 0$$

for all $\omega \in \Omega$ (see Theorem 3.1 and Corollary 3.1 again).

Furthermore, various optimality criteria which are not explicitly defined in terms of information matrices are proposed in the context of space-filling designs (e.g., [20]). For example, Johnson et al. [12] introduce optimality criteria related to the distance of observation points on hypercube, and discusses the relationship between D- or G-optimality criteria and the following two distance criteria.

- The maximin-distance criterion seeks for an n-point set X such that

$$X \text{ maximizes } \min_{i \neq j} \|x_i - x_j\|.$$

- The minimax-distance criterion seeks for an n-point set X such that

$$X \text{ minimizes } \max_x \min_{x_i} \|x - x_i\|.$$

In addition, there are several optimality criteria in terms of *potential energies*. A most comprehensive criteria is the Cohn–Kumar universal optimality criterion [5] for spherical configurations. Let $f : (0, 4] \to [0, \infty)$ be any decreasing and continuous function.

- The Cohn–Kumar universal optimality criterion seeks for an n-point set X such that

$$X \text{ minimizes the } f\text{-}potential\ energy \sum_{i \neq j} f(\|x_i - x_j\|^2)$$

for each completely monotonic potential function f.

In particular, when $f(u) = u^{-\alpha}$ for $\alpha > 0$, f-potential energy is known as the *discrete Riesz potential energy*. There are a number of works which examine "goodness" of spherical point configurations based on Riesz potential energy (e.g., [4, 10]). A somewhat challenging problem is to clarify the relationship between Riesz potential energy and some classical optimality criteria in statistics. For example, it may be interesting to find out whether Riesz potential minimization problem can be regarded as a dual problem of finding D-optimal designs, i.e., as a determinant maximization problem of information matrices.

References

1. Atkinson, A.C., Donev, A.N., Tobias, R.D.: Optimum Experimental Designs, With SAS, Oxford Statistical Science Series, vol. 34. Oxford University Press, Oxford (2007)
2. Bannai, E., Bannai, E.: On optimal tight 4-designs on 2 concentric spheres. Eur. J. Comb. **27**(2), 179–192 (2006)
3. Box, G.E.P., Draper, N.R.: Response Surfaces, Mixtures, and Ridge Analyses, 2nd edn. Wiley Series in Probability and Statistics. Wiley-Interscience, Wiley, Hoboken, NJ (2007)

4. Brauchart, J.S.: Optimal discrete Riesz energy and discrepancy. Unif. Distrib. Theory **6**(2), 207–220 (2011)
5. Cohn, H., Kumer, A.: Universally optimal distribution of points on spheres. J. Am. Math. Soc. **20**(1), 99–148 (2007)
6. Dean, A., Voss, D.: Design and Analysis of Experiments. Springer Texts in Statistics, 2nd edn. Springer, Cham (1999)
7. Dette, H., Melas, V.B., Pepelyshev, A.: Optimal designs for statistical analysis with Zernike polynomials. Statistics **41**(6), 453–470 (2007)
8. Draper, N., Pukelsheim, F.: On third order rotatability. Metrika **41**(1–2), 137–162 (1994)
9. Farrell, R.H., Kiefer, J., Walbran, A.: Optimum multivariate designs. In: Proceedings Fifth Berkeley Symposium, vol. 1, pp. 113–138. University California Press, Berkeley, Calif (1967)
10. Hardin, D.P., Michaels, T.J., Saff, E.: Asymptotic linear programming lower bounds for the energy of minimizing Riesz and Gauss configurations. Mathematika **65**(1), 157–180 (2019)
11. Hirao, M., Sawa, M., Jimbo, M.: Constructions of Φ_p-optimal rotatable designs on the ball. Sankhyā Ser A **77**(1), 211–236 (2015)
12. Johnson, M., Moore, L., Ylvisaker, D.: Minimax and maximin distance designs. J. Stat. Plan. Inference **26**, 131–148 (1990)
13. Karlin, S., Studden, W.J.: Tchebycheff Systems. Pure and Applied Mathematics, Vol. XV. Interscience Publishers, Wiley, New York-London-Sydney (1966)
14. Kiefer, J.: Optimum experimental designs. J. Roy. Stat. Soc. Ser. B **21**, 273–319 (1959)
15. Kiefer, J.: Optimum experimental designs V, with applications to systematic and rotatable designs. In: Proceedings 4th Berkeley Symposium, vol. 1, pp. 381–405. University California Press, Berkeley, Calif (1960)
16. Kiefer, J.: General equivalence theory for optimum designs (approximate theory). Ann. Stat. **2**, 849–879 (1974)
17. Kiefer, J.: Optimal design: variation in structure and performance under change of criterion. Biometrika **62**, 277–288 (1975)
18. Kiefer, J., Wolfowitz, J.: The equivalence of two extremum problems. Can. J. Math. **12**, 363–366 (1960)
19. Neumaier, A., Seidel, J.J.: Measures of strength $2e$ and optimal designs of degree e. Sankhyā Ser. A **54**, 299–309 (1992)
20. Pronzato, L., Müller, W.: Design of computer experiments: space filling and beyond. Stat. Comput. **22**(3), 681–701 (2012)
21. Pukelsheim, F.: Optimal Design of Experiments, Classics in Applied Mathematic, vol. 50. Society for Industrial and Applied Mathematics (SIAM), Philadelphia, PA (2006). Reprint of the 1993 original
22. Ramos-López, D., Sánchez-Granero, M.A., Fernandez-Martinez, M.: Optimal sampling patterns for zernike polynomials. Appl. Math. Comput. **274**, 247–257 (2016)
23. Rockafellar, R.T.: Convex Analysis. Princeton University Press, N.J. (1970)
24. Sawa, M., Hirao, M.: Characterizing D-optimal rotatable designs with finite reflection groups. Sankhyā Ser. A **79**(1), 101–132 (2017)
25. Silvey, S.D.: Optimal Design. Monographs on Applied Probability and Statistics. Chapman & Hall, London, New York (1980)

Chapter 4
Constructions of Optimal Euclidean Design

The most straightforward way of replacing an optimal invariant design ξ^* of degree e on unit ball $\mathbb{B}^d$ by a Euclidean $2e$-design (X, w) is to solve the moment equations $\sum_{x \in X} w(x) f(x) = \int_{\mathbb{B}^d} f(\omega) \xi^*(d\omega)$ for all monomials of degree up to $2e$. But a serious concern for this approach is that the number of equations rapidly grows with the number of unknown parameters θ_i even in the quadratic or cubic regression (3.1) in Sect. 3.1.

A commonly used approach is to restrict a weighted pair (X, w) to a class of invariant weighted pairs under a finite subgroup G of orthogonal group $\mathcal{O}(d)$ (cf. [27, 35]). It is often the case in design of experiments that G is taken to be hyperocta-hedral group B_d. A great advantage of such situations is the drastic reduction of the optimum design problem to an optimization problem with small steps involving only five or six moments (e.g., [12]). A series of previous works, among many others, are due to Gaffke and Heiligers [11–13] (see also [14]) who propose a good algorithm for numerically exploring optimal Euclidean designs (X, w) and thereby produce many examples. Their algorithm is based on techniques on convex analysis for square matrices; for a detailed account of those techniques, we refer the reader to Pukelsheim's book *Optimal Design of Experiments* [36].

The main purpose of the present chapter is to provide a method of explicitly constructing optimal Euclidean designs, called *corner vector method*. Intuitively, corner vectors are vertices, the midpoints of edges, the barycenter of faces, etc. of a regular polytope in $\mathbb{R}^d$. For example, when $G = B_d$, corner vectors are points of type $(1, \ldots, 1, 0, \ldots, 0)$ that have been traditionally employed in the construction theory of optimal Euclidean designs. The corner vector method provides a full generalization of this traditional construction of Euclidean designs, as well as various classical fractional factorial designs such as Scheffé $\{d + 1, 2\}$-lattice design [40], central composite designs [6], Box–Hunter polygonal designs [5], and so on. On the other hand, this chapter includes a different type of construction methods called *thinning method*. The idea is to first prepare a G-invariant Euclidean design and then

The original version of this chapter was revised: Missed out author corrections have been incorporated. The correction to this chapter is available at https://doi.org/10.1007/978-981-13-8075-4_6.

M. Sawa et al., *Euclidean Design Theory*, JSS Research Series in Statistics, https://doi.org/10.1007/978-981-13-8075-4_4

reduce the number of observation points through various statistical and combinatorial arrangements such as orthogonal arrays, combinatorial t-designs, and so on.

While Victoir [42] appears to be the first who formulates such methods in the context of the cubature theory, similar ideas can be found in some previous works in design of experiments; for example, see [4]. The thinning method described in the present chapter shows a full generalization of those previous works in numerical analysis and design of experiments.

Section 4.1 gives preliminaries, where definitions and basic properties on finite reflection groups are reviewed. Section 4.2 outlines the theory for invariant Euclidean designs, with particular emphasis on the Sobolev theorem [41] that is a key for further arguments. Section 4.3 discusses the corner vector method, especially, for three classical reflection groups A_d, B_d, D_d, and provides many examples with infinite families of D-optimal Euclidean designs. Section 4.4 first outlines the ideas of thinning method for group B_d and afterward, presents some important examples of families of D-optimal Euclidean designs by combining the corner vector method with the thinning method. Finally, Sect. 4.5 describes the conclusion including further remarks and open questions related.

4.1   Finite Reflection Groups

In this section, definitions and properties concerning finite reflection groups are reviewed. Most of the materials appearing in this section can be found in Broubaki [3], Coxeter [7], Humphreys [23], and references therein.

Given a nonzero vector $\alpha \in \mathbb{R}^d$, let $H_\alpha = \{\omega \in \mathbb{R}^d \mid \langle \omega, \alpha \rangle = 0\}$ be the hyperplane orthogonal to α. An orthogonal linear transformation $r_\alpha : \mathbb{R}^d \to \mathbb{R}^d$ is called a *reflection with respect to H_α* if

$$r_\alpha(\omega) = \omega - \frac{2\langle \omega, \alpha \rangle}{\langle \alpha, \alpha \rangle} \alpha \quad \text{for all } \omega \in \mathbb{R}^d.$$

A finite group generated by reflections is called a *finite reflection group*. Note that a finite reflection group is a subgroup of orthogonal group $\mathcal{O}(d)$ of $\mathbb{R}^d$ since

$$\langle r_\alpha(\omega), r_\alpha(\omega') \rangle = \langle \omega, \omega' \rangle \quad \text{for all } \omega, \omega' \in \mathbb{R}^d.$$

A *root system* Φ in $\mathbb{R}^d$ is a finite subset of $\mathbb{R}^d \setminus \{0\}$ satisfying the following two conditions:

(i) $\Phi \cap \mathrm{Span}_{\mathbb{R}}\{\alpha\} = \{\alpha, -\alpha\}$ for all $\alpha \in \Phi$.
(ii) $r_\alpha(\Phi) = \Phi$ for all $\alpha \in \Phi$ where $r_\alpha(\Phi) = \{r_\alpha(\omega) \mid \omega \in \Phi\}$.

Moreover, a subset Π of root system Φ in $\mathbb{R}^d$ is called a *fundamental system* if

(i) Π is linearly independent,
(ii) when $x \in \Phi$, $x = \sum_{y \in \Pi} c_y y$, where all $c_y \geq 0$ or $c_y \leq 0$.

It is well-known (e.g., [23, Chap. 2.4]) that fundamental systems are conjugate.

Given a root system Φ in $\mathbb{R}^d$, let G be the finite reflection group generated by all r_α, $\alpha \in \Phi$. Further, let $\Pi = \{\alpha_1, \ldots, \alpha_d\}$ be a fundamental system of Φ. Then $w = r_{\alpha_1} \cdots r_{\alpha_d}$ is called the *Coxeter element* of G. The *exponents* of G are defined to be the integers m of $0 < m \le h$ such that ξ^m is an eigenvalue of the Coxeter element of G in its representation on $\mathbb{R}^d$, where $\xi = \exp(-2\pi\sqrt{-1}/h)$ and h is the order of w.

The following concept plays a key role in the description of our method of constructing D-optimal Euclidean designs in Sect. 4.3.

Definition 4.1 (*Corner vector*) Let $\alpha_1, \ldots, \alpha_d \in \mathbb{R}^d$ be the fundamental roots of a finite reflection group. The *corner vectors* $v_1, \ldots, v_d \in \mathbb{R}^d$ are to satisfy that for every $i = 1, \ldots, d$, $v_i \perp \alpha_j$ if and only if $i \ne j$.

Given a fundamental system Π of root system Φ, the *Coxeter graph* is a graph with vertices Π, where two vertices α and β are joined by an edge if and only if the order of reflections $r_\alpha r_\beta$ is greater than or equal to 3. A finite reflection group G is said to be *irreducible* if and only if its Coxeter graph is connected. It is well-known that finite irreducible reflection groups are completely classified, as the following shows.

Theorem 4.1 (cf. [3, 7, 23]) *A finite irreducible reflection group G is classified by groups A_d $(d \ge 1)$, B_d $(d \ge 2)$, D_d $(d \ge 4)$ and exceptional groups H_3, H_4, F_4, E_6, E_7, E_8 and $I_2(n)$.*

In the subsequent part of this section, the certain information about finite irreducible reflection groups A_d, B_d and D_d such as fundamental roots, corner vectors, and exponents are briefly presented one by one, which will be used to construct D-optimal Euclidean designs in Sects. 4.2 and 4.3. See [3, 23] for the definition and some elementary facts about exceptional groups. Group $I_2(n)$ is the Dihedral group of order $2n$ acting on a regular n-gon in $\mathbb{R}^2$, which often appear in the present book.

4.1.1 Group A_d

Throughout this subsection, let $G = A_d$, $d \ge 1$. This group can be realized as the symmetry group of a regular simplex in $\mathbb{R}^d$ which is isomorphic to symmetric group $\mathscr{S}_{d+1}$ defined over a set of $d + 1$ symbols. Traditionally, a regular simplex is defined as

$$\mathscr{T}^{d+1} = \{\omega \in \mathbb{R}^{d+1} \mid \|\omega\|_1 = \omega_1 + \cdots + \omega_{d+1} = 1 \text{ and } \omega_i \ge 0 \text{ for every } i\}$$

in the usual context of the cubature theory; see e.g., Sawa and Xu [39]. But in this subsection and subsequent sections, deviating from such traditional manners, $\mathscr{T}^{d+1}$ is transformed to a certain solid regular simplex T^d embedded in $\mathbb{R}^d$ by combining projections with affine transformations (see Fig. 4.1). These two expressions of a

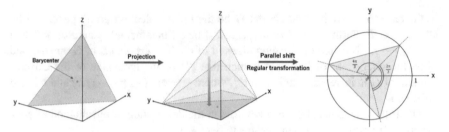

Fig. 4.1 Projection of $\mathscr{T}^3$ onto the xy-plane and affine transformation of the right-angled triangle to a solid regular simplex T^2

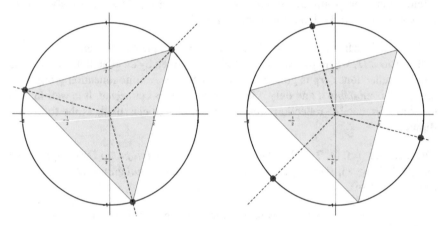

Fig. 4.2 Orbits $v_1^{A_2}$ (left) and $v_2^{A_2}$ (right)

simplex convey the same geometric meanings. However, the associated technicalities fairly differ, namely, the latter is inscribed in unit sphere $\mathbb{S}^{d-1}$ so as to be naturally handled in the setting of Theorem 3.3 for D-optimal invariant designs; see also Fig. 4.2. Most of the materials given below can be found in Nozaki and Sawa [33, pp. 8–12].[1]

Fundamental roots

$$\alpha_i = e_i - e_{i+1}, \quad 1 \le i \le d - 1,$$
$$\alpha_d = (a, a, \ldots, a, b)$$

with $a = (-1 + \sqrt{d+1})/d$ and $b = (d - 1 + \sqrt{d+1})/d$.

---

[1]Note that Broubaki [3] and Sali [37, pp. 163–164] deal with materials such as fundamental roots, reflections, corner vectors, and orbits, defined for regular simplex $\mathscr{T}^{d+1}$ but not for solid regular simplex T^d.

Reflections

The reflections r_{α_i} are regarded as elements of general linear group $GL_d(\mathbb{R})$ like

$$r_{\alpha_i} = (e_1^T, \ldots, e_{i-1}^T, e_{i+1}^T, e_i^T, e_{i+2}^T, \ldots, e_d^T), \quad 1 \le i \le d-1,$$

$$r_{\alpha_d} = \begin{pmatrix} E_{d-1} - a^2(\iota_{d-1}^T \iota_{d-1}) & -ab\iota_{d-1}^T \\ -ab\iota_{d-1} & 1 - b^2, \end{pmatrix}$$

where E_d is the identity matrix of order d, and ι_d is the all-one d-dimensional row vector.

Corner vectors

$$v_k = c_k \sum_{i=1}^{k} e_i + d_k \sum_{i=k+1}^{d} e_i,$$

where

$$c_k = \frac{d+1-k+\sqrt{d+1}}{\sqrt{k(d+1-k)(d+2+2\sqrt{d+1})}}, \quad d_k = \frac{-k}{\sqrt{k(d+1-k)(d+2+2\sqrt{d+1})}}.$$

Orbits[2]

Let $U_1 \subset \mathbb{R}^d$ be the set of all vectors with exactly k coordinates being c_k and the other $d-k$ coordinates d_k. Similarly, let $U_2 \subset \mathbb{R}^d$ be the set of all vectors with exactly $k-1$ coordinates being $\tilde{c}_k$ and the other coordinates $\tilde{d}_k$, where

$$\tilde{c}_k = \frac{d+1-k}{\sqrt{k(d+1-k)(d+2+2\sqrt{d+1})}}, \quad \tilde{d}_k = \frac{-k-\sqrt{d+1}}{\sqrt{k(d+1-k)(d+2+2\sqrt{d+1})}}.$$

Then it follows that

$$v_k^{A_d} = U_1 \cup U_2, \quad v_k^{A_d} = -v_{d+1-k}^{A_d}, \quad |v_k^{A_d}| = \binom{d+1}{k}.$$

For each k, the orbit $v_k^{A_d}$ corresponds to the set of barycenters of all k-dimensional faces of a regular simplex inscribed in $\mathbb{S}^{d-1}$.

Exponents

$1, 2, \ldots, d$.

Example 4.1 The group A_2 is generated by reflections of root system $\Phi = \left\{ \pm e_1 \mp \right.$

$e_2, \pm e_2, \pm \frac{-1+\sqrt{3}}{2} e_1 \pm \frac{1+\sqrt{3}}{2} e_2 \left. \right\}$ in $\mathbb{R}^2$. It is easy to check that

---

[2]Let G be a finite reflection group and v be a point on $\mathbb{S}^{d-1}$. Denote by v^G the orbit of v under G, i.e., $v^G = \{v^g \mid g \in G\}$.

$$\Pi = \left\{ \alpha_1 = e_1 - e_2, \alpha_2 = \frac{-1+\sqrt{3}}{2} e_1 + \frac{1+\sqrt{3}}{2} e_2 \right\}$$

is a fundamental system of Φ. Then reflections $r_{\alpha_1}, r_{\alpha_2}$ are regarded as elements of $GL_2(\mathbb{R})$ like

$$r_{\alpha_1} = \begin{pmatrix} 0 & -1 \\ 1 & 0 \end{pmatrix}, \quad r_{\alpha_2} = \begin{pmatrix} \frac{\sqrt{3}}{2} & -\frac{1}{2} \\ -\frac{1}{2} & -\frac{\sqrt{3}}{2} \end{pmatrix}.$$

The Coxeter element $r_{\alpha_1} r_{\alpha_2}$ of A_2 is represented as

$$\begin{pmatrix} 0 & -1 \\ 1 & 0 \end{pmatrix} \begin{pmatrix} \frac{\sqrt{3}}{2} & -\frac{1}{2} \\ -\frac{1}{2} & -\frac{\sqrt{3}}{2} \end{pmatrix} = \begin{pmatrix} \frac{1}{2} & \frac{\sqrt{3}}{2} \\ \frac{\sqrt{3}}{2} & -\frac{1}{2} \end{pmatrix}$$

being of order 2. Since $r_{\alpha_1} r_{\alpha_2}$ has the eigenvalues

$$-1 = \exp(-\pi\sqrt{-1}), \quad 1 = \exp(-2\pi\sqrt{-1}),$$

the exponents of A_2 are 1, 2.

Remark 4.1 The orbits $v_1^{A_2} \cup v_2^{A_2}$ form vertices of a regular 6-gon inscribed in $\mathbb{S}^1$; the orbit $v_1^{A_2}$ can be obtained by taking the point $((\sqrt{2}+\sqrt{6})/4, (\sqrt{2}-\sqrt{6})/4)$ and by rotating it by $2\pi/3$ and $4\pi/3$ radians. Similarly, $v_2^{A_2}$ can be obtained by taking the normalized midpoint $(1/\sqrt{2}, 1/\sqrt{2})$ and by rotating it by $2\pi/3$ and $4\pi/3$ radians; see Fig. 4.2. These six points have the same structure as the Hexagon design, which is one of the polygonal designs introduced by Box and Hunter [5].

In the general dimension case, the orbit $v_1^{A_d} \cup v_d^{A_d}$ is essentially the same as the Scheffé $\{d+1, 2\}$-lattice design on a regular simplex in the first orthant [40], whose D-optimality is discussed in [28]; see also Fig. 4.3.

4.1.2 Group B_d

In this subsection, let $G = B_d$, $d \geq 2$. This is the semi-direct product of $\mathscr{S}_d$ and $(\mathbb{Z}/2\mathbb{Z})^d$, where $\mathscr{S}_d$ consists of all reflections permuting e_i in $\mathbb{R}^d$ and $(\mathbb{Z}/2\mathbb{Z})^d$ consists of all coordinate-wise sign changes of e_i, which can be realized as the symmetry group of a hyperoctahedron or its dual hypercube in $\mathbb{R}^d$. It is then obvious that B_d has order $2^d d!$. Most of the materials given here can be found in Broubaki [3], Nozaki and Sawa [33, pp. 12–13], Sali [37, pp. 164–165], and references therein.

Fundamental roots

$$\alpha_i = e_i - e_{i+1}, \quad 1 \leq i \leq d-1, \quad \alpha_d = \sqrt{2} e_d.$$

Fig. 4.3 The Scheffé {3, 2}-lattice design

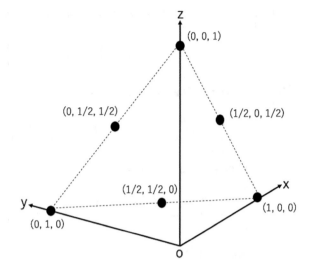

Corner vectors

$$v_i = \frac{1}{\sqrt{i}} \sum_{k=1}^{i} e_k, \quad 1 \le i \le d,$$

where each of v_i has exactly i nonzero coordinates.

Reflections

The reflections r_{α_i} are regarded as elements of general linear group $GL_d(\mathbb{R})$ like

$$r_{\alpha_i} = (e_1^T, \ldots, e_{i-1}^T, e_{i+1}^T, e_i^T, e_{i+2}^T, \ldots, e_d^T), \quad 1 \le i \le d-1,$$
$$r_{\alpha_d} = (e_1^T, \ldots, e_{d-1}^T, -e_d^T).$$

Orbits

The orbit $v_k^{B_d}$ is the set of vectors with exactly k nonzero coordinates equal to $\pm 1/\sqrt{k}$, and the size of $v_k^{B_d}$ (or the cardinality of $v_k^{B_d}$) is given by

$$|v_k^{B_d}| = 2^k \binom{d}{k}, \quad 1 \le k \le d.$$

Exponents

$1, 3, \ldots, 2d - 1$.

Example 4.2 The group B_2 is generated by reflections of root system $\Phi = \{\pm e_1 \pm e_2, \pm e_1 \mp e_2, \pm\sqrt{2}e_1, \pm\sqrt{2}e_2\}$ in $\mathbb{R}^2$. It is easy to check that

$$\Pi = \{\alpha_1 = e_1 - e_2, \alpha_2 = \sqrt{2}e_2\}$$

is a fundamental system of Φ. Then reflections $r_{\alpha_1}, r_{\alpha_2}$ are regarded as elements of $GL_2(\mathbb{R})$ like

$$r_{\alpha_1} = \begin{pmatrix} 0 & 1 \\ 1 & 0 \end{pmatrix}, \quad r_{\alpha_2} = \begin{pmatrix} 1 & 0 \\ 0 & -1 \end{pmatrix}.$$

The Coxeter element $r_{\alpha_1} r_{\alpha_2}$ is represented as

$$\begin{pmatrix} 1 & 0 \\ 0 & -1 \end{pmatrix} \begin{pmatrix} 0 & 1 \\ 1 & 0 \end{pmatrix} = \begin{pmatrix} 0 & 1 \\ -1 & 0 \end{pmatrix} = \begin{pmatrix} \cos(-\pi/2) & -\sin(-\pi/2) \\ \sin(-\pi/2) & \cos(-\pi/2) \end{pmatrix}$$

being of order 4. Since $r_{\alpha_1} r_{\alpha_2}$ has the eigenvalues

$$-\sqrt{-1} = \exp\left(-\frac{\pi\sqrt{-1}}{2}\right), \quad \sqrt{-1} = \exp\left(-\frac{3\pi\sqrt{-1}}{2}\right),$$

the exponents are 1, 3.

Remark 4.2 It is noted that $v_1^{B_d} \cup v_d^{B_d}$ to which origin 0 is added has the same structure as a central composite design, which is an important class of second-order response surface designs consisting of 2^d full factorials $(\pm 1, \ldots, \pm 1)$, star points $\pm e_i$ and central run 0. Also, totally, eight points in $v_1^{B_2} \cup v_2^{B_2}$ form a regular 8-gon inscribed in unit circle $\mathbb{S}^1$ as illustrated in Fig. 4.4, again belonging to an important class of polygonal designs (e.g., Box and Hunter [5]).

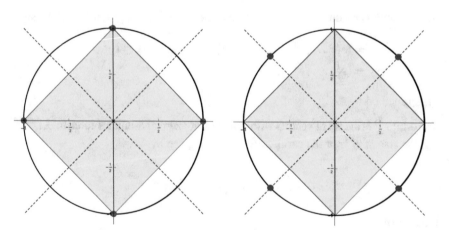

Fig. 4.4 Orbits $v_1^{B_2}$ (left) and $v_2^{B_2}$ (right)

4.1.3 Group D_d

Let G be group D_d for $d \geq 3$. This group is a two-index subgroup of group B_d, which is the symmetry group of a demihypercube in $\mathbb{R}^d$, i.e., semi-direct product of $\mathscr{S}_d$ and $(\mathbb{Z}/2\mathbb{Z})^{d-1}$ of order $2^{d-1}d!$. It is well-known (cf. Broubaki [3], Humphreys [23]) that group D_3 is isomorphic to group A_3. Most of the materials given here can be found in Broubaki [3], Nozaki and Sawa [33, pp. 14–16], Sali [37, p. 165], and references therein.

Fundamental roots

$$\alpha_i = e_i - e_{i+1}, \quad 1 \leq i \leq d - 1, \quad \alpha_d = e_{d-1} + e_d.$$

Reflections

The reflections r_{α_i} are regarded as elements of general linear group $GL_d(\mathbb{R})$ like

$$r_{\alpha_i} = (e_1^T, \ldots, e_{i-1}^T, e_{i+1}^T, e_i^T, e_{i+2}^T, \ldots, e_d^T), \quad 1 \leq i \leq d - 1,$$
$$r_{\alpha_d} = (e_1^T, \ldots, e_{d-2}^T, -e_d^T, -e_{d-1}^T).$$

Corner vectors

$$v_i = \frac{\sum_{k=1}^{i} e_k}{\sqrt{i}}, \quad 1 \leq i \leq d - 2, \quad v_{d-1} = \frac{\sum_{k=1}^{d-1} e_k - e_d}{\sqrt{d}}, \quad v_d = \frac{\sum_{k=1}^{d} e_k}{\sqrt{d}}.$$

Orbits

$v_k^{D_d} = v_k^{B_d}$ for $1 \leq k \leq d - 2$, and $v_{d-1}^{D_d}$ (respectively, $v_d^{D_d}$) consists of the vectors in $\{\pm 1/\sqrt{d}\}^n$ with an odd (respectively, even) number of coordinates of negative signs. Hence, $v_{d-1}^{D_d} \cup v_d^{D_d}$ partitions $v_d^{B_d}$. This means that if both $v_{d-1}^{D_d}$ and $v_d^{D_d}$ are employed, then the corner vector method for D_d can be regarded as a refinement of the corner vector method for B_d. The size of the D_d-orbit of v_i is given by

$$|v_i^{D_d}| = 2^i \binom{d}{i}, \quad 1 \leq i \leq d - 2, \quad |v_{d-1}^{D_d}| = |v_{d-1}^{D_d}| = 2^{d-1}.$$

Exponents

$1, 3, \ldots, 2d - 3, d - 1.$

Example 4.3 The group D_3 is generated by root system $\Phi = \{\pm e_1 \pm e_2, \pm e_1 \mp e_2 \pm e_2 \pm e_3, \pm e_2 \mp e_3, \pm e_1 \pm e_3, \pm e_1 \mp e_3\}$ in $\mathbb{R}^6$. It is easy to check that

$$\Pi = \{\alpha_1 = e_1 - e_2, \alpha_2 = e_2 - e_3, \alpha_3 = e_2 + e_3\}$$

is a fundamental system of Φ. Then reflections $r_{\alpha_1}, r_{\alpha_2}, r_{\alpha_3}$ are regarded as elements of $GL_3(\mathbb{R})$ like

$$
r_{\alpha_1} = \begin{pmatrix} 0 & 1 & 0 \\ 1 & 0 & 0 \\ 0 & 0 & 1 \end{pmatrix}, \quad
r_{\alpha_2} = \begin{pmatrix} 1 & 0 & 0 \\ 0 & 0 & 1 \\ 0 & 1 & 0 \end{pmatrix}, \quad
r_{\alpha_3} = \begin{pmatrix} 1 & 0 & 0 \\ 0 & 0 & -1 \\ 0 & 1 & 0 \end{pmatrix}.
$$

The Coxeter element $r_{\alpha_1} r_{\alpha_2} r_{\alpha_3}$ of D_3 is represented as

$$
\begin{pmatrix} 0 & 1 & 0 \\ 1 & 0 & 0 \\ 0 & 0 & 1 \end{pmatrix}
\begin{pmatrix} 1 & 0 & 0 \\ 0 & 0 & 1 \\ 0 & 1 & 0 \end{pmatrix}
\begin{pmatrix} 1 & 0 & 0 \\ 0 & 0 & -1 \\ 0 & 1 & 0 \end{pmatrix}
= \begin{pmatrix} 0 & 1 & 0 \\ 1 & 0 & 0 \\ 0 & 0 & -1 \end{pmatrix}
$$

being of order 2. Since $r_{\alpha_1} r_{\alpha_2} r_{\alpha_3}$ has the eigenvalues

$$
-1 = \exp\left(-\pi\sqrt{-1}\right), \quad 1 = \exp\left(-2\pi\sqrt{-1}\right), \quad -1 = \exp\left(-3\pi\sqrt{-1}\right),
$$

the exponents are $1, 2, 3$.

Remark 4.3 The orbits $v_1^{D_3}$, $v_2^{D_3}$, and $v_3^{D_3}$ are given as follows:

$$
v_1^{D_3} = \left\{ \frac{1}{\sqrt{3}}(\pm 1, 0, 0), \frac{1}{\sqrt{3}}(0, \pm 1, 0), \frac{1}{\sqrt{3}}(0, 0, \pm 1) \right\},
$$

$$
v_2^{D_3} = \left\{ \frac{1}{\sqrt{3}}(1, 1, -1), \frac{1}{\sqrt{3}}(1, -1, 1), \frac{1}{\sqrt{3}}(-1, 1, 1), \frac{1}{\sqrt{3}}(-1, -1, -1) \right\},
$$

$$
v_3^{D_3} = \left\{ \frac{1}{\sqrt{3}}(1, -1, -1), \frac{1}{\sqrt{3}}(-1, -1, 1), \frac{1}{\sqrt{3}}(-1, 1, -1), \frac{1}{\sqrt{3}}(1, 1, 1) \right\}.
$$

Namely, orbit $v_1^{D_3}$ forms six vertices of a regular octahedron, and orbits $v_2^{D_3}$ and $v_3^{D_3}$ form two partitions of barycentres of eight faces of the octahedron; see Fig. 4.5.

In statistics language, vectors $v_1^{D_3} \cup v_2^{D_3} \cup v_3^{D_3}$ together with origin 0 have the same structure as a three-factor central composite design.

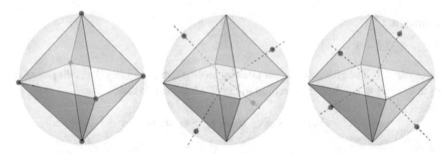

Fig. 4.5 Orbits $v_1^{D_3}$ (left), $v_2^{D_3}$ (middle), and $v_3^{D_3}$ (right)

4.2 Invariant Polynomial and the Sobolev Theorem

The purpose of this section is to give a detailed explanation of the Sobolev theorem [41], which plays a key role in the development of corner vector methods as will be described in the next section.

At first, the definition of invariant polynomials is given.

Let $G \subset \mathcal{O}(d)$ be a finite reflection group. The action of $g \in G$ on $f \in \mathscr{P}(\mathbb{R}^d)$ is defined by

$$f^g(\omega) = f(\omega^{g^{-1}}), \qquad \omega \in \mathbb{R}^d.$$

A polynomial f is said to be *G-invariant* if $f^g = f$ for every $g \in G$. Denote by $\mathscr{P}_t(\mathbb{R}^d)^G$ and $\mathrm{Harm}_t(\mathbb{R}^d)^G$, the set of G-invariant polynomials in $\mathscr{P}_t(\mathbb{R}^d)$ and $\mathrm{Harm}_t(\mathbb{R}^d)$, respectively.

Definition 4.2 (*Invariant weighted pair*) A finite weighted pair (X, w) is said to be *G-invariant* if

(i) X is a union of G-orbits,
(ii) for each $x \in X$, $w(y) = w(y')$ for every $y, y' \in x^G$.

The following fact is known as the Sobolev theorem.

Theorem 4.2 (Sobolev theorem, [41]) *Let (X, w) be a G-invariant finite weighted pair. Then the following are equivalent:*

(i) *A pair (X, w) is a Euclidean t-design supported by unit sphere $\mathbb{S}^{d-1}$, i.e., it holds that*

$$\frac{1}{|\mathbb{S}^{d-1}|} \int_{\mathbb{S}^{d-1}} f(\omega)\, \rho(d\omega) = \sum_{x \in X} w(x) f(x) \quad \text{for all } f \in \mathscr{P}_t(\mathbb{R}^d).$$

(ii) *It holds that*

$$\frac{1}{|\mathbb{S}^{d-1}|} \int_{\mathbb{S}^{d-1}} f(\omega)\, \rho(d\omega) = \sum_{x \in X} w(x) f(x) \quad \text{for all } f \in \mathscr{P}_t(\mathbb{R}^d)^G.$$

Example 4.4 Let us apply Theorem 4.2 to show that

$$\frac{1}{|\mathbb{S}^1|} \int_{\mathbb{S}^1} f(\omega)\, \rho(d\omega) = \frac{1}{4} \sum_{x \in \{(\pm 1, \pm 1)\}} f(x) \quad \text{for all } f \in \mathscr{P}_3(\mathbb{S}^1).$$

Take $G = \left\langle \begin{pmatrix} 1 & 0 \\ 0 & -1 \end{pmatrix}, \begin{pmatrix} 0 & -1 \\ 1 & 0 \end{pmatrix} \right\rangle$, which occasionally coincides with group B_2 and so has exponents 1 and 3, as already seen in Example 4.2. As will be seen in Theorem 4.4,

$\mathscr{P}_3(\mathbb{S}^1)^G$ is spanned by the constant function $f \equiv 1$. The fact that vertices of a regular 4-gon form a spherical (tight) 3-design has already been introduced in Example 2.8.

Next, note that G is isomorphic to Dihedral group $I_2(4)$ of order 8. It then follows from standard calculations that any G-invariant bivariate polynomial on $\mathbb{S}^1$ is expressed by a polynomial of $\cos 4\theta$ in polar coordinate system $(x, y) = (\cos\theta, \sin\theta)$. Since $\cos 4\theta$ corresponds to homogeneous polynomial $g(x, y) = x^4 - 6x^2y^2 + y^4$ by trigonometric addition formulas, there do not exist nontrivial G-invariant polynomials of degrees 2 and 3. In summary, $\mathscr{P}_4(\mathbb{S}^1)^G$ is spanned by the constant function $f \equiv 1$ and the above g. Thus, by checking the definition property (2.5) only for the above f and g, it is easily seen that the vertices of a regular 5-gon form a (tight) spherical 4-design.

Nozaki and Sawa [33] generalize the Sobolev theorem for Euclidean designs supported by a single sphere to multiple concentric spheres; see also Theorem 2.8 in Sect. 2.3.

Theorem 4.3 (Sobolev theorem for Euclidean design, [33]) *Let (X, w) be a G-invariant weighted pair with $X = \bigcup_k (r_k x_k)^G$, where $x_k \in \mathbb{S}^{d-1}$ and $r_k > 0$. The following are equivalent:*

(i) A pair (X, w) is a G-invariant Euclidean t-design.
(ii) $\sum_{x \in X} w(x) \|x\|^{2j} \phi(x) = 0$ for every $\phi \in \mathrm{Harm}_\ell(\mathbb{R}^d)^G$ with $1 \le \ell \le t, 0 \le j \le \lfloor \frac{t-\ell}{2} \rfloor$.

A proof is here provided to make the present book more self-contained. For the proof, the following lemma is first needed.

Lemma 4.1 *Let G be a subgroup of $\mathcal{O}(d)$. Further let f be a G-invariant polynomial and x^G be a G-orbit. Then it holds that $f(y) = f(x)$ for all $y \in x^G$.*

Proof The definition of a G-invariant polynomial shows that for given $g \in G$

$$f(y) = f(x^g) = f^{g^{-1}}(x) = f(x). \qquad \square$$

Proof of Theorem 4.3. For $f \in \mathrm{Harm}_\ell(\mathbb{R}^d)$, it is seen that the polynomial

$$\phi(\xi) = \frac{1}{|G|} \sum_{g \in G} f(\xi^g)$$

is an element of $\mathrm{Harm}_\ell(\mathbb{R}^d)^G$; this is so-called *Reynolds operator* (see Sect. 5.5). Let $w(x) = w_k$ for every $x \in (r_k x_k)^G$. Then it follows from Lemma 4.1 that for every $f \in \mathrm{Harm}_\ell(\mathbb{R}^d)$

$$\sum_{x \in X} w(x)\|x\|^{2j} f(x) = \sum_k w_k r_k^{2j} \sum_{x \in x_k^G} f(r_k x)$$

$$= \sum_k \frac{w_k r_k^{2j+\ell} |x_k^G|}{|G|} \sum_{g \in G} f(x_k^g)$$

$$= \sum_k w_k r_k^{2j+\ell} |x_k^G| \phi(x_k)$$

$$= \sum_k w_k r_k^{2j} \sum_{x \in x_k^G} \phi(r_k x)$$

$$= \sum_{x \in X} w(x)\|x\|^{2j} \phi(x).$$

The required result thus follows from Theorem 2.8. $\qquad\square$

On account of Theorem 4.3, Corollary 4.1 can be obtained.

Corollary 4.1 *Let (X, w) be a G-invariant finite weighted pair. A pair (X, w) is a G-invariant Euclidean t-design supported by unit sphere $\mathbb{S}^{d-1}$ if and only if it holds that $\sum_{x \in X} w(x)\|x\|^{2j}\phi(x) = 0$ for every $\phi \in \mathrm{Harm}_\ell(\mathbb{R}^d)^G$ with $1 \le \ell \le t, 0 \le j \le \lfloor \frac{t-\ell}{2} \rfloor$.*

Since the space of G-invariant harmonic polynomials has very small dimension, Theorems 4.2 and 4.3 drastically reduce the computational cost of solving a system of algebraic equations involving the defining property (2.7) of Euclidean designs.

Let $q_i = \dim \mathrm{Harm}_i(\mathbb{R}^d)^G$. Then q_i is explicitly calculated by use of the *harmonic Molien-Poincaré series*.

Theorem 4.4 ([15]) *Let $G \subset \mathcal{O}(d)$ be a finite (irreducible) reflection group. Then*

$$\sum_{i=0}^{\infty} q_i u^i = \prod_{i=2}^{d} \frac{1}{1 - u^{1+n_i}}$$

where $1 = n_1 \le n_2 \le \cdots \le n_d$ are the exponents of G (see Sect. 4.1 for the definition of exponent).

In the subsequent part of this section, harmonic Molien-Poincaré series, dimension of invariant harmonic polynomial space and some related materials are computed for groups A_d ($d \ge 2$), B_d ($d \ge 2$), D_d ($d \ge 4$). It is convenient to use the notation $\mathrm{sym}(f)$ for a symmetric polynomial as defined by

$$\mathrm{sym}(f) := \frac{1}{|(\mathscr{S}_d)_f|} \sum_{g \in \mathscr{S}_d} f(x^g)$$

with $(\mathscr{S}_d)_f := \{g \in \mathscr{S}_d \mid f(x^g) = f(x) \text{ for all } x \in \mathbb{R}^d\}$. For example, when $f(x) = x_1^2 \in \mathscr{P}_2(\mathbb{R}^3)$, it holds that

$$\mathrm{sym}(x_1^2) = \frac{1}{2}(2x_1^2 + 2x_2^2 + 2x_3^2) = x_1^2 + x_2^2 + x_3^2$$

since $|(\mathscr{S}_3)_{x_1^2}| = |\{e, (2, 3)\}| = 2$.

4.2.1   Group A_d

Harmonic Molien-Poincaré series

With the exponents of A_d described in Sect. 4.1.1, the harmonic Molien-Poincaré series is given by

$$\frac{1}{(1 - u^3)(1 - u^4) \cdots (1 - u^{d+1})} = \begin{cases} 1 + u^3 + u^6 + \mathscr{O}(u^9), & \text{if } d = 2, \\ 1 + u^3 + u^4 + u^6 + \mathscr{O}(u^7), & \text{if } d = 3, \\ 1 + u^3 + u^4 + u^5 + \mathscr{O}(u^6), & \text{if } d \geq 4. \end{cases}$$

Dimensions of G-invariant harmonic polynomial spaces

By combining the above harmonic Molien-Poincaré series and Theorem 4.4, dimensions of A_d-invariant harmonic polynomial spaces are given by

$$\dim \mathrm{Harm}_1(\mathbb{R}^d)^{A_d} = \dim \mathrm{Harm}_2(\mathbb{R}^d)^{A_d} = 0, \quad d \geq 2,$$
$$\dim \mathrm{Harm}_3(\mathbb{R}^d)^{A_d} = 1, \quad d \geq 2,$$
$$\dim \mathrm{Harm}_4(\mathbb{R}^d)^{A_d} = \begin{cases} 0, & d = 2, \\ 1, & d \geq 3, \end{cases}$$
$$\dim \mathrm{Harm}_5(\mathbb{R}^d)^{D_d} = \begin{cases} 0, & d = 2, 3, \\ 1, & d \geq 4. \end{cases}$$

Basis of G-invariant harmonic polynomial spaces

By direct calculations, it is easy to find a basis of A_d-invariant harmonic polynomial spaces.

• The space $\mathrm{Harm}_3(\mathbb{R}^d)^{A_d}$ is spanned by the following f_3 as

$$f_3 = \begin{cases} x_1^3 - 3x_1^2 x_2 - 3x_1 x_2^2 + x_2^3, & d = 2, \\ \mathrm{sym}(x_1^3) - \frac{3}{2}\mathrm{sym}(x_1 x_2^2) - \frac{3}{4}\mathrm{sym}(x_1 x_2 x_3), & d = 3, \\ \mathrm{sym}(x_1^3) - \frac{3}{d-1}\mathrm{sym}(x_1 x_2^2) + \frac{6(2 - \sqrt{d+1})}{(d-1)(d-3)}\mathrm{sym}(x_1 x_2 x_3), & d \geq 4. \end{cases}$$

• The space $\mathrm{Harm}_4(\mathbb{R}^d)^{A_d}$ is spanned by the following f_4 as

$$f_4 = \begin{cases} h_{4,1} - \dfrac{20}{13}h_{4,3}, & d = 3, \\ h_{4,1} + c_{4,2}h_{4,2} + c_{4,3}f_{4,3}, & d \geq 4, \end{cases}$$

where

$$h_{4,1} = \mathrm{sym}(x_1^4) - \frac{6}{d-2}\mathrm{sym}(x_1^2 x_2^2),$$

$$h_{4,2} = \mathrm{sym}(x_1 x_2 x_3 x_4),$$

$$h_{4,3} = \mathrm{sym}(x_1 x_2^3) - \frac{6}{d-2}\mathrm{sym}(x_1 x_2 x_3^2)$$

and

$$c_{4,2} = \frac{24(d+2)(d^2 - 5d - 12 + 4\sqrt{d+1})}{(d-1)(d-2)(d^3 - 2d^2 - 15d - 16)},$$

$$c_{4,3} = -\frac{4(d+2)(d^2 - 2d - 7 - (d-1)\sqrt{d+1})}{(d-1)(d^3 - 2d^2 - 15d - 16)}.$$

- The space $\mathrm{Harm}_5(\mathbb{R}^d)^{A_d}$ is spanned by the following f_5 as

$$f_5 = \begin{cases} h_{5,1} + \dfrac{17 - 20\sqrt{5}}{58}h_{5,2} + \dfrac{10(18 + \sqrt{5})}{87}h_{5,3}, & d = 4, \\ h_{5,1} + c_{5,2}h_{5,2} + c_{5,3}f_{5,3} + c_{5,4}h_{5,4}, & d \geq 4, \end{cases}$$

where

$$h_{5,1} = \mathrm{sym}(x_1^5) - \frac{10}{d-1}\mathrm{sym}(x_1^2 x_2^3) + \frac{30}{(d-1)(d-2)}\mathrm{sym}(x_1 x_2^2 x_3^2),$$

$$h_{5,2} = \mathrm{sym}(x_1^5) - \frac{10}{d-1}\mathrm{sym}(x_1^2 x_2^3) + \frac{5}{d-1}\mathrm{sym}(x_1 x_2^4),$$

$$h_{5,3} = \mathrm{sym}(x_1 x_2 x_3^3) - \frac{9}{d-3}\mathrm{sym}(x_1 x_2 x_3 x_4^2),$$

$$h_{5,4} = \mathrm{sym}(x_1 x_2 x_3 x_4 x_5)$$

and

$$c_{5,2} = -\frac{2d^3 + 5d^2 - 21d - 90 - d(d+6)\sqrt{d+1}}{4d^3 + 3d^2 - 60d - 180},$$

$$c_{5,3} = \frac{20(2d^3 + 6d^2 - 32d - 168 + (d^2 - 8d + 12)\sqrt{d+1}}{(d-1)(d-2)(4d^3 + 3d^2 - 60d - 180)},$$

$$c_{5,4} = -\frac{120(d+6)(d^2 - 11d - 78 + (2d^2 - 2d + 12)\sqrt{d+1})}{(d-1)(d-2)(d-3)(4d^3 + 3d^2 - 60d - 180)}.$$

Substituting v_k for G-invariant harmonic polynomials

1. Degree 3.

   (i) For $d \geq 2, d \neq 3$

   $$f_3(v_k) = -\frac{k - \frac{d+1}{2}}{\sqrt{k(d+1-k)}}\phi_3(d),$$

   where

   $$\phi_3(d) = \frac{2(d^3 + 3d^2 - 12d - 16 + (3d^2 - 4d - 16)\sqrt{d+1})}{(d-1)(d-3)(d+2+2\sqrt{d+1})^{3/2}}.$$

   (ii) For $d = 3$

   $$f_3(v_1) = \frac{3\sqrt{3}}{4}, \quad f_3(v_2) = 0, \quad f_3(v_3) = -\frac{3\sqrt{3}}{4}.$$

2. Degree 4.
   For $d \geq 3$

   $$f_4(v_k) = \frac{(k-\alpha)(k-\beta)}{k(d+1-k)}\phi_4(d)$$

   with

   $$\phi_4(d) = \frac{6(d+1)(d^5 + 7d^4 - 24d^3 - 160d^2 - 256d - 128 + 4(d^4 - 20d^2 - 28d - 32)\sqrt{d+1})}{(d-1)(d-2)(d^3 - 2d^2 - 15d - 16)(d+2+2\sqrt{d+1})^2},$$

   $$\alpha = \frac{d+1}{2} - \frac{\sqrt{3(d^2-1)}}{6}, \quad \beta = \frac{d+1}{2} + \frac{\sqrt{3(d^2-1)}}{6}.$$

3. Degree 5.
   For $d \geq 4$

   $$f_5(v_k) = -\frac{(k - \frac{d+1}{2})(k - \alpha')(k - \beta')}{\{k(d+1-k)\}^{3/2}}\phi_5(d)$$

   with

   $$\phi_5(d) = \frac{24(d+1)(2d^6 + 31d^5 + 50d^4 - 448d^3 - 2144d^2 - 3200d - 1536)}{(d-1)(d-2)(d-3)(4d^3 + 3d^2 - 60d - 180)(d+2+2\sqrt{d+1})^{3/2}}$$
   $$+ \frac{24(d+1)(11d^5 + 50d^4 - 96d^3 - 1120d^2 - 2432d - 1536)\sqrt{d+1}}{(d-1)(d-2)(d-3)(4d^3 + 3d^2 - 60d - 180)(d+2+2\sqrt{d+1})^{3/2}},$$

   $$\alpha' = \frac{d+1}{2} - \frac{\sqrt{3(d+1)(2d-3)}}{6}, \quad \beta' = \frac{d+1}{2} + \frac{\sqrt{3(d+1)(2d-3)}}{6}.$$

4.2.2 Group B_d

Harmonic Molien-Poincaré series

With the exponents of B_d described in Sect. 4.1.2, the harmonic Molien-Poincaré series is given by

$$\frac{1}{(1-u^4)(1-u^6)\cdots(1-u^{2d})} = \begin{cases} 1 + u^4 + u^8 + \mathcal{O}(u^{12}), & \text{if } d = 2, \\ 1 + u^4 + u^6 + u^8 + \mathcal{O}(u^{10}), & \text{if } d = 3, \\ 1 + u^4 + u^6 + 2u^8 + \mathcal{O}(u^{10}), & \text{if } d \geq 4. \end{cases}$$

Dimensions of G-invariant harmonic polynomial spaces

By combining the above harmonic Molien-Poincaré series and Theorem 4.4, dimensions of B_d-invariant harmonic polynomial spaces are given by

$$\dim \text{Harm}_1(\mathbb{R}^d)^{B_d} = \dim \text{Harm}_2(\mathbb{R}^d)^{B_d} = \dim \text{Harm}_3(\mathbb{R}^d)^{B_d} = 0, \quad d \geq 2,$$

$$\dim \text{Harm}_4(\mathbb{R}^d)^{B_d} = 1, \quad d \geq 2,$$

$$\dim \text{Harm}_6(\mathbb{R}^d)^{B_d} = \begin{cases} 0, & d = 2, \\ 1, & d \geq 3. \end{cases}$$

$$\dim \text{Harm}_8(\mathbb{R}^d)^{B_d} = \begin{cases} 1, & d = 2, 3, \\ 2, & d = 4. \end{cases}$$

Basis of G-invariant harmonic polynomial spaces

By direct calculations, it is easy to find a basis of B_d-invariant harmonic polynomial spaces.
- The space $\text{Harm}_4(\mathbb{R}^d)^{B_d}$ is spanned by the following polynomials f_4 as

$$f_4 = \text{sym}(x_1^4) - \frac{6}{d-1}\text{sym}(x_1^2 x_2^2), \quad d \geq 2.$$

- The space $\text{Harm}_6(\mathbb{R}^d)^{B_d}$ is spanned by the following polynomials f_6 as

$$f_6 = \text{sym}(x_1^6) - \frac{15}{d-1}\text{sym}(x_1^4 x_2^2) + \frac{180}{(d-1)(d-2)}\text{sym}(x_1^2 x_2^2 x_3^2), \quad d \geq 3.$$

- The space $\text{Harm}_8(\mathbb{R}^d)^{B_d}$ is spanned by the following polynomials $f_{8,1}$ and $f_{8,2}$, respectively, as

$$f_{8,1} = \text{sym}(x_1^8) - \frac{28}{d-1}\text{sym}(x_1^2 x_2^6) + \frac{70}{d-1}\text{sym}(x_1^4 x_2^4), \quad d \geq 2,$$

$$f_{8,2} = \text{sym}(x_1^4 x_2^4) - \frac{6}{d-2}\text{sym}(x_1^2 x_2^4 x_3^2)$$

$$+ \frac{108}{(d-2)(d-3)}\text{sym}(x_1^2 x_2^2 x_3^2 x_4^2), \quad d \geq 4.$$

Substituting v_k for G-invariant harmonic polynomials

1. Degree 4.
   For $d \geq 2$

   $$f_4(v_k) = \frac{1}{k}\left(1 - 3\frac{k-1}{d-1}\right).$$

2. Degree 6.
   For $d \geq 3$

   $$f_6(v_k) = \frac{1}{k^2}\left(1 - 15\frac{k-1}{d-1} + 30\frac{(k-1)(k-2)}{(d-1)(d-2)}\right).$$

3. Degree 8.

   (i) For $d \geq 2$

   $$f_{8,1}(v_k) = \frac{1}{k^3}\left(1 + 7\frac{k-1}{d-1}\right).$$

   (ii) For $d \geq 4$

   $$f_{8,2}(v_k) = \frac{k-1}{2k^3}\left(1 - 6\frac{k-2}{d-2} + 9\frac{(k-2)(k-3)}{(d-2)(d-3)}\right).$$

4.2.3   Group D_d

Harmonic Molien-Poincaré series

With the exponents of D_d described in Sect. 4.1.3, the harmonic Molien-Poincaré series is given by

$$\frac{1}{(1-u^4)(1-u^6)\cdots(1-u^{2d-2})(1-u^d)} = \begin{cases} 1 + 2u^4 + u^6 + 3u^8 + \mathcal{O}(u^{10}), & \text{if } d = 4, \\ 1 + u^4 + u^5 + u^6 + 2u^8 + \mathcal{O}(u^9), & \text{if } d = 5, \\ 1 + u^4 + 2u^6 + 2u^8 + \mathcal{O}(u^{10}), & \text{if } d = 6, \\ 1 + u^4 + u^6 + u^7 + 2u^8 + \mathcal{O}(u^{10}), & \text{if } d = 7, \\ 1 + u^4 + u^6 + 3u^8 + \mathcal{O}(u^{10}), & \text{if } d = 8, \\ 1 + u^4 + u^6 + 2u^8 + \mathcal{O}(u^{10}), & \text{if } d \geq 9. \end{cases}$$

Dimensions of G-invariant harmonic polynomial spaces

By combining the above harmonic Molien-Poincaré series and Theorem 4.4, dimensions of D_d-invariant harmonic polynomial spaces are given by

$$\dim \text{Harm}_1(\mathbb{R}^d)^{D_d} = \dim \text{Harm}_2(\mathbb{R}^d)^{D_d} = \dim \text{Harm}_3(\mathbb{R}^d)^{D_d} = 0, \quad d \geq 4,$$

$$\dim \text{Harm}_4(\mathbb{R}^d)^{D_d} = \begin{cases} 2, & d = 4, \\ 1, & d \geq 5, \end{cases}$$

$$\dim \mathrm{Harm}_5(\mathbb{R}^d)^{D_d} = \begin{cases} 0, & d = 4, \ d \geq 6, \\ 1, & d = 5, \end{cases}$$

$$\dim \mathrm{Harm}_6(\mathbb{R}^d)^{D_d} = \begin{cases} 1, & d = 4, 5, \ d \geq 7, \\ 2, & d = 6, \end{cases}$$

$$\dim \mathrm{Harm}_7(\mathbb{R}^d)^{D_d} = \begin{cases} 0, & d = 4, 5, 6, \ d \geq 8, \\ 1, & d = 7, \end{cases}$$

$$\dim \mathrm{Harm}_8(\mathbb{R}^d)^{D_d} = \begin{cases} 3, & d = 4, 8 \\ 2, & d \geq 5, d \neq 8. \end{cases}$$

Basis of G-invariant harmonic polynomial spaces

By direct calculations, it is easy to find a basis of D_d-invariant harmonic polynomial spaces.

- The space $\mathrm{Harm}_4(\mathbb{R}^d)^{D_d}$ is spanned by the following polynomials $f_{4,1}$ and $f_{4,2}$ as

$$f_{4,1}(x) = \mathrm{sym}(x_1^4) - \frac{6}{d-1}\mathrm{sym}(x_1^2 x_2^2), \quad d \geq 5,$$

$$f_{4,2}(x) = x_1 x_2 x_3 x_4, \quad d = 4.$$

- The space $\mathrm{Harm}_5(\mathbb{R}^5)^{D_5}$ is spanned by the following f_5 as

$$f_5 = x_1 x_2 x_3 x_4 x_5.$$

- $\mathrm{Harm}_6(\mathbb{R}^d)^{D_d}$ is spanned by the following polynomials $f_{6,1}$ and $f_{6,2}$ as

$$f_{6,1} = \mathrm{sym}(x_1^6) - \frac{15}{d-1}\mathrm{sym}(x_1^2 x_2^4) + \frac{180}{(d-1)(d-2)}\mathrm{sym}(x_1^2 x_2^2 x_3^2), \quad d \geq 4,$$

$$f_{6,2} = x_1 x_2 x_3 x_4 x_5 x_6, \quad d = 6.$$

- $\mathrm{Harm}_7(\mathbb{R}^7)^{D_7}$ is spanned by

$$f_7 = x_1 x_2 x_3 x_4 x_5 x_6 x_7.$$

- The space $\mathrm{Harm}_8(\mathbb{R}^d)^{B_d}$ is spanned by the following polynomials $f_{8,1}$, $f_{8,2}$, $f_{8,3}$ and $f_{8,4}$ respectively, as

$$f_{8,1} = \mathrm{sym}(x_1^8) - \frac{28}{d-1}\mathrm{sym}(x_1^2 x_2^6) + \frac{70}{d-1}\mathrm{sym}(x_1^4 x_2^4), \quad d \geq 4,$$

$$f_{8,2} = \mathrm{sym}(x_1^4 x_2^4) - \frac{6}{d-2}\mathrm{sym}(x_1^2 x_2^2 x_3^4)$$

$$+ \frac{108}{(d-2)(d-3)}\mathrm{sym}(x_1^2 x_2^2 x_3^2 x_4^2), \quad d \geq 4,$$

$$f_{8,3} = \mathrm{sym}(x_1^5 x_2 x_3 x_4) - \frac{10}{9}\mathrm{sym}(x_1^3 x_2^3 x_3 x_4), \quad d = 4.$$

For $d = 8$

$$f_{8,4} = x_1 x_2 x_3 x_4 x_5 x_6 x_7 x_8.$$

Substituting v_k for G-invariant harmonic polynomials

1. Degree 4.
   (i) For $d \geq 5$

$$
f_{4,1}(v_k) =
\begin{cases}
\dfrac{1}{k}\left(1 - 3\dfrac{k-1}{d-1}\right), & 1 \leq k \leq d-2, \\[2mm]
-\dfrac{2}{d}, & k = d-1, d.
\end{cases}
$$

   (ii) For $d = 4$

$$f_{4,2}(v_1) = f_{4,2}(v_2) = 0, \quad f_{4,2}(v_3) = -\frac{1}{16}, \quad f_{4,2}(v_4) = \frac{1}{16}.$$

2. Degree 5.
   For $d = 5$

$$f_5(v_1) = f_5(v_2) = f_5(v_3) = 0, \quad f_5(v_4) = -\frac{1}{25\sqrt{5}}, \quad f_5(v_5) = \frac{1}{25\sqrt{5}}.$$

3. Degree 6.
   (i) For $d \geq 4$

$$
f_{6,1}(v_k) =
\begin{cases}
\dfrac{1}{k^2}\left(1 - 15\dfrac{k-1}{d-1} + 30\dfrac{(k-1)(k-2)}{(d-1)(d-2)}\right), & 1 \leq k \leq d-2, \\[2mm]
\dfrac{16}{d^2}, & k = d-1, d.
\end{cases}
$$

   (ii) For $d = 6$

$$f_{6,2}(v_1) = \cdots = f_{6,2}(v_4) = 0, \quad f_{6,2}(v_5) = -\frac{1}{216}, \quad f_{6,2}(v_6) = \frac{1}{216}.$$

4. Degree 7.
   For $d = 7$

$$f_7(v_1) = \cdots = f_7(v_5) = 0, \quad f_7(v_6) = -\frac{1}{343\sqrt{7}}, \quad f(v_7) = \frac{1}{343\sqrt{7}}.$$

5. Degree 8.
   (i) For $d \geq 4$

$$
f_{8,1}(v_k) =
\begin{cases}
\dfrac{1}{k^3}\left(1 + 7\dfrac{k-1}{d-1}\right), & 1 \leq k \leq d-2, \\[2mm]
\dfrac{8}{d^3}, & k = d-1, d.
\end{cases}
$$

   (ii) For $d \geq 4$

$$
f_{8,2}(v_k) = \begin{cases} \dfrac{k-1}{2k^3}\left(1 - 6\dfrac{k-2}{d-2} + 9\dfrac{(k-2)(k-3)}{(d-2)(d-3)}\right), & 1 \le k \le d-2, \\ \dfrac{2(d-1)}{d^3}, & k = d-1, d. \end{cases}
$$

(iii) For $d = 4$

$$
f_{8,3}(v_1) = f_{8,3}(v_2) = 0, \quad f_{8,3}(v_3) = \frac{1}{96}, \quad f_{8,3}(v_4) = -\frac{1}{96}.
$$

(iv) For $d = 8$

$$
f_{8,4}(v_1) = \cdots = f_{8,4}(v_6) = 0, \quad f_{8,4}(v_7) = -\frac{1}{4096}, \quad f_{8,4}(v_8) = \frac{1}{4096}.
$$

4.3 Corner Vector Methods

As already seen several times in the previous sections, corner vectors associated with a regular simplex or a hyperoctahedron in $\mathbb{R}^d$ provide a geometric understanding of various classes of experimental designs such as Scheffé lattice design [40], Box–Behnken design [4], Box–Hunter polygonal design [5], and so on. Then a natural direction of research is toward the application of corner vectors to the construction of optimal Euclidean designs. In this section a powerful method for constructing optimal Euclidean designs, called *corner vector method*, is provided together with a number of infinite families and individual examples of optimal designs. Among many other optimality criteria, our attention is restricted to D-optimality for simplicity of description. It is also assumed that experimental domains under consideration are unit ball $\mathbb{B}^d$, but there are no technical differences to obtain similar results for other experimental domains such as hypercubes, the entire space $\mathbb{R}^d$, and so on.

Let G be a finite reflection group that can be realized as the symmetry group of a regular polytope in $\mathbb{R}^d$, namely, G is taken to be one of the groups classified in Theorem 4.1. Let $\mathcal{R}$ be the radius set of $\lfloor e/2 \rfloor + 1$ concentric spheres $\mathbb{S}^{d-1}_{r_1}, \ldots, \mathbb{S}^{d-1}_{r_{\lfloor e/2 \rfloor + 1}}$. It is assumed without loss of generality that $r_1 > \cdots > r_{\lfloor e/2 \rfloor + 1} \ge 0$; for convenience let $r_{\lfloor e/2 \rfloor + 1} = 0$ if e is an even integer. Further, let $\mathcal{J} = \{J_1, \ldots, J_{\lfloor e/2 \rfloor + 1}\}$ be a subset of power set $2^{\{1,2,\ldots,d\}}$. For convenience, when $r_{\lfloor e/2 \rfloor + 1} = 0$, we set $J_{\lfloor e/2 \rfloor + 1} = \emptyset$. A conclusion emerging from Definition 3.3, Theorems 4.2 and 4.3 is the following method of constructing optimal designs.

Theorem 4.5 (Corner vector construction) *Let*

$$
\mathcal{X}(G, \mathcal{R}, \mathcal{J}) = \bigcup_{i=1}^{\lfloor e/2 \rfloor + 1} \bigcup_{k \in J_i} (r_i v_k)^G
$$

be a G-invariant set on $\lfloor e/2 \rfloor + 1$ concentric spheres. Especially, when $J_{\lfloor e/2 \rfloor + 1} = \emptyset$, i.e., e is even, we set $\bigcup_{k \in J_{\lfloor e/2 \rfloor + 1}} (r_{\lfloor e/2 \rfloor + 1} v_k)^G = \{0\}$. Further, let w be a G-invariant weight function on $\mathcal{X}(G, \mathcal{R}, \mathcal{J})$ satisfying

$$
w(x) \equiv w_k^{(i)} \text{ (constant)} \quad \text{for every } x \in (r_i v_k)^G.
$$

Assume that a pair $(\mathscr{R}, \{W_i\}_{i=1,...,\lfloor e/2 \rfloor+1})$ *is the unique solution of the optimization problem to maximize (3.13) in Sect. 3.3, where* $W_i = \sum_{x \in \bigcup_{k \in J_i} (r_i v_k)^G} w(x)$ *on* $\mathbb{S}_{r_i}^{d-1}$ *for each* i. *Then the following are equivalent:*

(i) $(\mathscr{X}(G, \mathscr{R}, \mathscr{J}), w)$ *is a D-optimal Euclidean 2e-design supported by* $\lfloor e/2 \rfloor + 1$ *concentric spheres.*

(ii) *There exists a solution of a system of equations given by*

$$\sum_{x \in \mathscr{X}(G, \mathscr{R}, \mathscr{J})} w(x) \|x\|^{2j} \phi(x) = 0$$

for every $\phi \in \mathrm{Harm}_\ell(\mathbb{R}^d)^G$ *with* $1 \le \ell \le 2e$, $0 \le j \le \lfloor \frac{2e-\ell}{2} \rfloor$.

Remark 4.4 As mentioned at the first part of this chapter, the simplest way of finding a D-optimal Euclidean 2e-design (X, w) is to solve the nonlinear moment equations $\sum_{x \in X} w(x) f(x) = \int f(\omega) \xi(d\omega)$ for all monomials f of degree up to $2e$. Such approaches have been traditionally used in design of experiments [4, 21]. However, they often suffer the serious disadvantage of heavy computational cost even for small values of t. For example, see Box–Behnken [4, Appendix A] and [32]. A great advantage of our approach is that the existence of a D-optimal Euclidean 2e-design can be reduced to finding a solution of a certain system of only a few nonlinear equations for *any* e. This will be explained in detail only for $e = 2$ and 3 in subsequent Sects. 4.3.1, 4.3.2, and 4.3.3, but is always true even for larger values of e. A concrete example for $e = 4$ is given in Yamamoto et al. [43, Theorem 2.5 and Remark 2.6]; see also Sawa and Hirao [38] for higher cases of e.

The following theorems (Theorems 4.6 and 4.7) mention the maximum degree $2e$ for which a G-invariant Euclidean design possibly exists.

Theorem 4.6 ([33]) *Let* G *be one of finite reflection groups* A_d, B_d, *and* D_d. *Further let*

(i) $e \ge 3$ *if* $G = A_d$,   (ii) $e \ge 4$ *if* $G = B_d$,   (iii) $e \ge 4$ *if* $G = D_d$.

Then there is no choice of $\mathscr{R}$, $\mathscr{J}$ *and w for which* $(\mathscr{X}(G, \mathscr{R}, \mathscr{J}), w)$ *is a Euclidean 2e-design supported by finitely many concentric spheres in* $\mathbb{R}^d$.

The evaluation of e described in Theorem 4.6 is "best", as the following Theorem 4.7 shows.

Theorem 4.7 ([33]) *Let* G *be one of finite reflection groups* A_d, B_d, *and* D_d. *Further let*

(i) $e = 2$ *if* $G = A_d$,   (ii) $e = 3$ *if* $G = B_d$,   (iii) $e = 3$ *if* $G = D_d$.

Then there exists a choice of $\mathscr{R}$, $\mathscr{J}$ *and w for which* $(\mathscr{X}(D_d, \mathscr{R}, \mathscr{J}), w)$ *is a Euclidean 2e-design supported by finitely many concentric spheres in* $\mathbb{R}^d$.

The following corollary follows immediately from Theorem 4.7.

Corollary 4.2 ([33]) *Let G be one of finite reflection groups A_d, B_d, and D_d. Further let*

(i) $e \geq 3$ if $G = A_d$,   (ii) $e \geq 4$ if $G = B_d$,   (iii) $e \geq 4$ if $G = D_d$.

Then there exists no choice of $\mathcal{R}$, $\mathcal{J}$ and w for which $(\mathcal{X}(G, \mathcal{R}, \mathcal{J}), w)$ is a D-optimal Euclidean 2e-design supported by finitely many concentric spheres in $\mathbb{R}^d$.

In the next subsection, it will be shown that the evaluations of values of e in Corollary 4.2 are best for $G = A_d, B_d, D_d$.

Note that most of the results given in the next subsection can be also found in Hirao et al. [20] and Sawa and Hirao [38].

4.3.1 Group A_d

By Corollary 4.2 (i), there does not exist a D-optimal Euclidean 6-design of type $(\mathcal{X}(A_d, \mathcal{R}, \mathcal{J}), w)$. Here, a D-optimal Euclidean 4-design of type $(\mathcal{X}(A_d, \mathcal{R}, \mathcal{J}), w)$ supported by $\mathbb{S}^{d-1} \cup \{0\}$ is presented, since it is sufficient to construct a D-optimal Euclidean 4-design when we consider the second-order regression model. Hereinafter, recalling (3.14), let $\mathcal{R} = \{1, 0\}$, $\mathcal{J} = \{J_1, \emptyset\}$ with a non-empty subset J_1 of $\{1, \ldots, d\}$, and let

$$W_1 = \sum_{k \in J_1} \sum_{x \in V_k^{A_d}} w(x) = \frac{d(d+3)}{(d+1)(d+2)}, \quad W_2 = w(0) = \frac{2}{(d+1)(d+2)}.$$

The following is the main result in this section.

Theorem 4.8 ([38], Theorem 9) *Let $d \geq 4$ and J_1 be a non-empty subset of $\{1, \ldots, d\}$. A pair $(\mathcal{X}(A_d, \{1, 0\}, \{J_1, \emptyset\}), w)$ is a D-optimal Euclidean 4-design supported by $\mathbb{S}^{d-1} \cup \{0\}$ if and only if there is a solution of a system of equations given by*

$$\sum_{k \in J_1} \frac{w_k^{(1)} \binom{d+1}{k} (k - \frac{d+1}{2})}{\sqrt{k(d+1-k)}} = 0, \tag{4.1}$$

$$\sum_{k \in J_1} \frac{w_k^{(1)} \binom{d+1}{k} (k - \alpha)(k - \beta)}{k(d+1-k)} = 0, \tag{4.2}$$

$$\sum_{k \in J_1} w_k^{(1)} \binom{d+1}{k} = \frac{d(d+3)}{(d+1)(d+2)}, \tag{4.3}$$

where

$$\alpha = \frac{d+1}{2} - \frac{\sqrt{3(d^2-1)}}{6}, \quad \beta = \frac{d+1}{2} + \frac{\sqrt{3(d^2-1)}}{6}.$$

Proof Let $d \geq 4$. By recalling Sect. 4.2.1,

$$\dim \mathrm{Harm}_1(\mathbb{R}^d)^{A_d} = \dim \mathrm{Harm}_2(\mathbb{R}^d)^{A_d} = 0.$$

Moreover, $\mathrm{Harm}_3(\mathbb{R}^d)^{A_d}$ and $\mathrm{Harm}_4(\mathbb{R}^d)^{A_d}$ are spanned by f_3 and f_4, respectively. By substituting each corner vector v_k into f_3 and f_4, it holds that

$$f_3(v_k) = -\frac{k - \frac{d+1}{2}}{\sqrt{k(d+1-k)}}\phi_3(d), \quad f_4(v_k) = \frac{(k-\alpha)(k-\beta)}{k(d+1-k)}\phi_4(d).$$

By applying Theorem 4.5, a pair $(\mathscr{X}(A_d, \{1, 0\}, \{J_1, \emptyset\}), w)$ is a Euclidean 4-design supported by $\mathbb{S}^{d-1} \cup \{0\}$ if and only if for $i = 3, 4$

$$0 = \sum_{x \in \mathscr{X}(A_d, \{1,0\})} w(x) f_i(x) = \sum_{k \in J_1} \sum_{x \in v_k^{A_d}} w_k^{(1)} f_i(x) = \sum_{k \in J_1} w_k^{(1)} \binom{d+1}{k} f_i(v_k),$$

where the last equality follows from Lemma 4.1 and the fact that $|v_k^{A_d}| = \binom{d+1}{k}$; see Sect. 4.1.1. These equations are equivalent to (4.1) and (4.2) since $\phi_3(d), \phi_4(d) \neq 0$ for $d \geq 4$. Finally, (4.3) just follows from the definition of W_1, which completes the proof. □

Theorem 4.8 yields many families of D-optimal Euclidean 4-designs.

Proposition 4.1 ([38], Proposition 2)

(i) *Let $d \geq 5$ be an odd integer and k be a positive integer such that*

$$1 \leq k \leq \frac{1+d}{2} - \frac{\sqrt{-1+d^2}}{2\sqrt{3}} \quad or \quad \frac{1+d}{2} + \frac{\sqrt{-1+d^2}}{2\sqrt{3}} \leq k \leq d.$$

Let $J_1 = \{k, \frac{d+1}{2}, d-k+1\}$. Then $(\mathscr{X}(A_d, \{1, 0\}, \{J_1, \emptyset\}), w)$ with

$$w_k^{(1)} = w_{d-k+1}^{(1)} = \frac{1}{\binom{d+1}{k}} \frac{k(d+1-k)(d-1)d(d+3)}{(d+1)(d+2)^2(1-2k+d)^2},$$

$$w_{\frac{d+1}{2}}^{(1)} = \frac{1}{\binom{d+1}{(d+1)/2}} \frac{d(d+3)(2-6k+6k^2+3d-6kd+d^2)}{(d+2)^2(1-2k+d)^2}$$

forms a D-optimal Euclidean 4-design.

(ii) Let d, k be positive integers such that $d \geq 4$ and

$$\frac{3 + 3d - \sqrt{3(-1 + d^2)}}{6} \leq k \leq \frac{1 + d - \sqrt{\frac{d(1+d)}{4+d}}}{2} \quad or$$

$$\frac{1 + d + \sqrt{\frac{d(1+d)}{4+d}}}{2} \leq k \leq \frac{3 + 3d + \sqrt{3(-1 + d^2)}}{6}.$$

Let $J_1 = \{k, d - k + 1, d\}$. Then $(\mathcal{X}(A_d, \{1, 0\}, \{J_1, \emptyset\}), w)$ with

$$w_k^{(1)} = w_{d-k+1}^{(1)} = \frac{1}{\binom{d+1}{k}} \frac{d(d+3)g(d, k)}{2(-1 + k)(k - d)(-1 + 2k - d)(d + 1)^2(d + 2)^2},$$

$$w_d^{(1)} = \frac{d^2(d+3)(2 - 6k + 6k^2 + 3d - 6kd + d^2)}{(-1 + k)(k - d)(d + 1)^3(d + 2)^2}$$

forms a Euclidean 4-design, where

$$g(d, k) = (d - 1)\{k(k - d - 1)(2k - d - 1)(d - 2)$$
$$- (2 - 6k + 6k^2 + 3d - 6kd + d^2)\sqrt{dk(1 - k + d)}\}.$$

Remark 4.5 Among conditions (4.1) through (4.3) in Theorem 4.8, the first two are substantial and the remaining one is just a side condition because it can be automatically achieved from the other two. Theorem 4.8 states that the existence of a D-optimal Euclidean 4-design is equivalent to the existence of a solution of a certain system of *only two* nonlinear equations, which drastically reduces computational cost in comparison with the approach of directly solving equations $\sum_i W_i \int_{\mathbb{S}_{r_i}^{d-1}} f(\omega)\rho_{r_i}(d\omega)/|\mathbb{S}_{r_i}^{d-1}| = \sum_{x \in X} w(x)f(x)$ for all monomials f of degree up to $2e$. This is also true for groups B_d and D_d. For example, when $G = B_d$, it is shown in Theorem 4.9 that the existence of a D-optimal Euclidean 6-design is equivalent to the existence of a solution of a certain system of two nonlinear equations.

4.3.2 Group B_d

By Corollary 4.2 (ii), there does not exist a D-optimal Euclidean 8-design of type $(\mathcal{X}(B_d, \mathcal{R}, \mathcal{J}), w)$. Here, a D-optimal Euclidean 6-design of type $(\mathcal{X}(B_d, \mathcal{R}, \mathcal{J}), w)$ supported by $\mathbb{S}^{d-1} \cup \mathbb{S}_r^{d-1}$ is presented. Hereinafter, let $\mathcal{R} = \{1, r\}$, $\mathcal{J} = \{J_1, J_2\}$ with non-empty subsets J_1, J_2 of $\{1, \ldots, d\}$, and let

$$W_1 = \sum_{k \in J_1} \sum_{x \in v_k^{B_d}} w(x) = 1 - W, \quad W_2 = \sum_{k \in J_2} \sum_{x \in (rv_k)^{B_d}} w(x) = W,$$

where a pair (r, W) is a solution of a system of equations (3.15).

The following is the main result in this subsection.

Theorem 4.9 ([20], Proposition 1) *Let $d \geq 3$ and J_1, J_2 be non-empty subsets of $\{1, \ldots, d\}$. A pair $(\mathscr{X}(B_d, \mathscr{R}, \{J_1, J_2\}), w)$ is a D-optimal Euclidean 6-design supported by $\mathbb{S}^{d-1} \cup \mathbb{S}^{d-1}_r$ if and only if there exists a solution of a system of equations given by*

$$0 = \sum_{k \in J_1} w_k^{(1)} \frac{2^{k+1}}{k^2} \binom{d-1}{k-1} \left(1 - 3\frac{k-1}{d-1}\right) = \sum_{k \in J_2} w_k^{(2)} \frac{2^{k+1}}{k^2} \binom{d-1}{k-1} \left(1 - 3\frac{k-1}{d-1}\right),$$
(4.4)

$$0 = \sum_{k \in J_1} 3w_k^{(1)} \frac{2^{k+1}}{k^3} \binom{d-1}{k-1} \left(1 - 15\frac{k-1}{d-1} + 30\frac{(k-1)(k-2)}{(d-1)(d-2)}\right)$$
$$+ \sum_{k \in J_2} 3w_k^{(2)} r^6 \frac{2^{k+1}}{k^3} \binom{d-1}{k-1} \left(1 - 15\frac{k-1}{d-1} + 30\frac{(k-1)(k-2)}{(d-1)(d-2)}\right),$$
(4.5)

$$W = 1 - \sum_{k \in J_1} w_k^{(1)} 2^k \binom{d}{k} = \sum_{k \in J_2} w_k^{(2)} 2^k \binom{d}{k}.$$
(4.6)

Proof Let $d \geq 3$. By recalling Sect. 4.2.2, it is seen that

$$\dim \mathrm{Harm}_i(\mathbb{R}^d)^{B_d} = 0, \quad i = 1, 2, 3, 5.$$

Moreover, $\mathrm{Harm}_4(\mathbb{R}^d)^{B_d}$ and $\mathrm{Harm}_6(\mathbb{R}^d)^{B_d}$ are spanned by f_4 and f_6, respectively. By substituting each corner vector v_k into f_4 and f_6, it holds that

$$f_4(v_k) = \frac{1}{k}\left(1 - 3\frac{k-1}{d-1}\right),$$
$$f_6(v_k) = \frac{1}{k^2}\left(1 - 15\frac{k-1}{d-1} + 30\frac{(k-1)(k-2)}{(d-1)(d-2)}\right).$$

By applying Theorem 4.5, a pair $(\mathscr{X}(B_d, \{1, r\}, \{J_1, J_2\}), w)$ is a Euclidean 6-design supported by $\mathbb{S}^{d-1} \cup \mathbb{S}^{d-1}_r$ if and only if

$$0 = \sum_{x \in \mathscr{X}(B_d, \{1,r\})} w(x)\|x\|^{2j} f_i(x)$$
$$= \sum_{k \in J_1} \sum_{x \in v_k^{B_d}} w_k^{(1)} f_i(x) + \sum_{k \in J_2} \sum_{x \in (rv_k)^{B_d}} w_k^{(2)} r^{2j} f_i(x)$$
$$= \sum_{k \in J_1} 2^k \binom{d}{k} w_k^{(1)} f_i(v_k) + \sum_{k \in J_2} 2^k \binom{d}{k} w_k^{(2)} r^{i+2j} f_i(v_k)$$
(4.7)

for $(i, j) = (4, 0), (4, 1), (6, 0)$, where the last equality follows from Lemma 4.1 and the fact that $|v_k^{B_d}| = 2^k \binom{d}{k}$; see Sect. 4.1.2. Since $0 < r < 1$, the above Eq. (4.7) for $(i, j) = (4, 0), (4, 1)$ are equivalent to (4.4). The above Eq. (4.7) for $(i, j) = (6, 0)$ is equivalent to (4.5). Finally, (4.6) just follows from the definition of W, which completes the proof. $\square$

The following proposition characterizes D-optimal Euclidean 6-designs with $|J_1| = 3$ and $|J_2| = 2$. For all positive integers k_1, k_2, let

$$G(k_1, k_2) := (d + 2 - 3k_1)(d + 2 - 3k_2) + 6(k_1 - 1)(k_2 - 1) + 2(d - 1).$$

Proposition 4.2 ([20], Proposition 5) *Let $J_1 = \{i_1, i_2, i_3\}$ and $J_2 = \{j_1, j_2\}$ such that $i_1 < i_2 < i_3$ with $i_1 < \frac{d+2}{3} < i_3$ and $j_1 < \frac{d+2}{3} < j_2$. Assume that*

$$j_1 j_2 G(i_2, i_3)(1 - W) + i_2 i_3 G(j_1, j_2) r^6 W > 0,$$
$$j_1 j_2 G(i_1, i_3)(1 - W) + i_1 i_3 G(j_1, j_2) r^6 W < 0,$$
$$j_1 j_2 G(i_1, i_2)(1 - W) + i_1 i_2 G(j_1, j_2) r^6 W > 0.$$

Then $(\mathscr{X}(B_d, \{1, r\}, \{J_1, J_2\}), w)$ with

$$w_{i_1}^{(1)} = \frac{i_1^2 \{j_1 j_2 G(i_2, i_3)(1 - W) + i_2 i_3 G(j_1, j_2) r^6 W\} i_1!(d - i_1)!}{2^{i_1}(i_2 - i_1)(i_3 - i_1) j_1 j_2 (d + 2)(d + 4) d!},$$

$$w_{i_2}^{(1)} = -\frac{i_2^2 \{j_1 j_2 G(i_1, i_3)(1 - W) + i_1 i_3 G(j_1, j_2) r^6 W\} i_2!(d - i_2)!}{2^{i_2}(i_2 - i_1)(i_3 - i_2) j_1 j_2 (d + 2)(d + 4) d!},$$

$$w_{i_3}^{(1)} = \frac{i_3^2 \{j_1 j_2 G(i_1, i_2)(1 - W) + i_1 i_2 G(j_1, j_2) r^6 W\} i_3!(d - i_3)!}{2^{i_3}(i_3 - i_1)(i_3 - i_2) j_1 j_2 (d + 2)(d + 4) d!},$$

$$w_{j_1}^{(2)} = \frac{j_1(d + 2 - 3j_2) W j_1!(d - j_1)!}{2^{j_1}(j_1 - j_2)(d + 2) d!}, \quad w_{j_2}^{(2)} = \frac{j_2(d + 2 - 3j_1) W j_2!(d - j_2)!}{2^{j_2}(j_2 - j_1)(d + 2) d!}.$$

forms a D-optimal Euclidean 6-design.

Proof The result follows from (4.4), (4.5), and (4.6). □

Example 4.5 Many examples of D-optimal Euclidean 6-designs with $|J_1| = 3$ and $|J_2| = 2$ are provided by Proposition 4.2. For the case $(d, i_1, i_2, i_3, j_1, j_2) = (4, 1, 2, 4, 1, 3)$, by combining Proposition 4.2 and Table 3.1 in Sect. 3.3, $(\mathscr{X}(B_4, \{1, 0.537239\}, \{\{1, 2, 4\}, \{1, 3\}\}), w)$ with

$$w_1^{(1)} = 0.0177159, \quad w_2^{(1)} = 0.0177655, \quad w_4^{(1)} = 0.0177159,$$
$$w_1^{(2)} = 0.00463894, \quad w_3^{(2)} = 0.0034792$$

forms a D-optimal Euclidean 6-design. Moreover, for the case $(d, i_1, i_2, i_3, j_1, j_2) = (7, 1, 3, 7, 1, 6)$, by combining Proposition 4.2 and Table 3.1 in Sect. 3.3, $(\mathscr{X}(B_7, \{1, 0.548728\}, \{\{1, 3, 7\}, \{1, 6\}\}), w)$ with

$$w_1^{(1)} = 0.00468281, \quad w_2^{(1)} = 0.00226964, \quad w_4^{(1)} = 0.00179264,$$
$$w_1^{(2)} = 0.000992644, \quad w_3^{(2)} = 0.00012408$$

forms a D-optimal Euclidean 6-design. Note that these optimal values are rounded to no more than six significant figures for convenience.

Next, by use of Corollary 4.1, Euclidean 6-designs supported by $\mathbb{S}^{d-1}$ of type $(\mathscr{X}(B_d, \{1\}, \{J_1\}), \tilde{w})$ are presented. Namely

$$\frac{1}{|\mathbb{S}^{d-1}|} \int_{\mathbb{S}^{d-1}} f(\omega)\, \rho(d\omega) = \sum_{x \in \mathscr{X}(B_d, \{1\}, \{J_1\})} \tilde{w}(x) f(x) \quad \text{for all } f \in \mathscr{P}_6(\mathbb{R}^d).$$

Similarly to the argument used in Proposition 4.2, the following lemma is obtained.

Lemma 4.2 ([20], Lemma 1) *Let* $J_1 = \{k, l\}$ *such that* $k < \frac{d+2}{3} < l$ *with* $l = \frac{(d+2-3k)(d+4)}{3(d+4-5k)}$. *Then* $(\mathscr{X}(B_d, \{1\}, \{J_1\}), \tilde{w})$ *with*

$$\tilde{w}_k^{(1)} = \frac{k(3l - d - 2)k!(d - k)!}{2^k(l - k)(d + 2)d!}, \quad \tilde{w}_l^{(1)} = \frac{l(d + 2 - 3k)l!(d - l)!}{2^l(l - k)(d + 2)d!}$$

forms a Euclidean 6-design supported by unit sphere $\mathbb{S}^{d-1}$.

Example 4.6 Let $(d, k) = (3m + 2, 1)$ with $m \geq 1$. By Lemma 4.2 (i), $(\mathscr{X}(B_{3m+2}, \{1\}, \{1, m + 2\}), \tilde{w})$ with

$$\tilde{w}_1^{(1)} = \frac{(1 + 3m)!}{(1 + m)(4 + 3m)(2 + 3m)!}, \quad \tilde{w}_1^{(m+2)} = \frac{(2 + m)(1 + 3m)(2m)!(2 + m)!}{2^{2+m}(1 + m)(4 + 3m)(2 + 3m)!}$$

forms a Euclidean 6-design supported by unit sphere $\mathbb{S}^{3m+1}$.

By combining these four Euclidean 6-designs supported by $\mathbb{S}^d$ and Proposition 2.1 in Sect. 2.3, many examples of Euclidean 6-designs supported by $\mathbb{S}^{d-1} \cup \mathbb{S}_r^{d-1}$ of type $(\mathscr{X}(B_d, \{1, r\}, \{J_1, J_2\}), w)$ are produced. Moreover, some infinite families of D-optimal Euclidean 6-designs are obtained.

Theorem 4.10 ([20], Theorem 11) *Let* $d = 3m + 2$ *with* $m \geq 1$. *Further, let* $J_1 = \{1, m + 2\}$ *and* $J_2 = \{1, m + 2\}$. *Then there exists a D-optimal Euclidean 6-design of type* $(\mathscr{X}(B_d, \{1, r\}, \{J_1, J_2\}), w)$ *supported by* $\mathbb{S}^{d-1} \cup \mathbb{S}_r^{d-1}$.

Proof Take two copies of a Euclidean 6-design of types $(\mathscr{X}(B_d, \{1\}, \{1, m + 2\}), w)$ on unit sphere $\mathbb{S}^{3m+1}$. Applying Proposition 2.1 to the optimal radius and weight (solutions of (3.15) in Sect. 3.3) provides a sequence of D-optimal Euclidean 6-designs of type $(\mathscr{X}(B_d, \{1, r\}, \{J_1, J_2\}), w)$ with

$$w(x) = \begin{cases} (1 - W)\tilde{w}_1^{(1)} = \dfrac{(1 - W)(1 + 3m)!}{(1 + m)(4 + 3m)(2 + 3m)!}, & x \in v_1^{B_{3m+2}}, \\[2mm] (1 - W)\tilde{w}_{m+2}^{(1)} = \dfrac{(1 - W)(2 + m)(1 + 3m)(2m)!(2 + m)!}{2^{2+m}(1 + m)(4 + 3m)(2 + 3m)!}, & x \in v_{m+2}^{B_{3m+2}}, \\[2mm] W\tilde{w}_1^{(1)} = \dfrac{W(1 + 3m)!}{(1 + m)(4 + 3m)(2 + 3m)!}, & x \in rv_1^{B_{3m+2}}, \\[2mm] W\tilde{w}_{m+2}^{(1)} = \dfrac{W(2 + m)(1 + 3m)(2m)!(2 + m)!}{2^{2+m}(1 + m)(4 + 3m)(2 + 3m)!}, & x \in rv_{m+2}^{B_{3m+2}}. \end{cases}$$

$\square$

4.3.3 Group D_d

By Corollary 4.2 (iii), there does not exist a D-optimal Euclidean 8-design of type $(\mathscr{X}(D_d, \mathscr{R}, \mathscr{J}), w)$. Here, a D-optimal Euclidean 6-design of type $(\mathscr{X}(D_d, \mathscr{R}, \mathscr{J}), w)$ supported by $\mathbb{S}^{d-1} \cup \mathbb{S}_r^{d-1}$ is presented. Hereinafter, let $\mathscr{R} = \{1, r\}$, $\mathscr{J} = \{J_1, J_2\}$ with non-empty subsets J_1, J_2 of $\{1, \ldots, d\}$, and let

$$W_1 = \sum_{k \in J_1} \sum_{x \in v_k^{B_d}} w(x) = 1 - W, \quad W_2 = \sum_{k \in J_2} \sum_{x \in (rv_k)^{B_d}} w(x) = W$$

where a pair (r, W) is a solution of a system of equations (3.15).

The following is the main theorem in this subsection.

Theorem 4.11 ([38], Theorem 8) *Let $d \geq 7$ and J_1, J_2 be non-empty subsets of $\{1, \ldots, d\}$. A pair $(\mathscr{X}(D_d, \{1, r\}, \{J_1, J_2\}), w)$ is a D-optimal Euclidean 6-design supported by $\mathbb{S}^{d-1} \cup \mathbb{S}_r^{d-1}$ if and only if there exists a solution of a system of equations given by*

$$0 = \sum_{k \in J_i \setminus \{d-1, d\}} \frac{w_k^{(i)} 2^k \binom{d}{k} \left(1 - 3\frac{k-1}{d-1}\right)}{k} - \frac{\delta_{d-1, J_i} 2^d w_{d-1}^{(i)}}{d} - \frac{\delta_{d, J_i} 2^d w_d^{(i)}}{d}, \quad i = 1, 2,$$

(4.8)

$$0 = \sum_{i=1}^{2} \sum_{k \in J_i \setminus \{d-1, d\}} r_i^6 \left\{ \frac{w_k^{(i)} 2^k \binom{d}{k}}{k^2} \left(1 - 15\frac{k-1}{d-1} + 30\frac{(k-1)(k-2)}{(d-1)(d-2)}\right) \right.$$

$$\left. + \delta_{d-1, J_i} \frac{2^{d+3} w_{d-1}^{(i)}}{d^2} + \delta_{d, J_i} \frac{2^{d+3} w_d^{(i)}}{n^2} \right\},$$

(4.9)

$$W_1 = \sum_{k \in J_1 \setminus \{d-1, d\}} w_k^{(1)} 2^k \binom{d}{k} + \delta_{d-1, J_1} 2^{d-1} w_{d-1}^{(1)} + \delta_{d, J_1} 2^{d-1} w_d^{(1)},$$

(4.10)

$$W_2 = \sum_{k \in J_2 \setminus \{d-1, d\}} w_k^{(2)} 2^k \binom{d}{k} + \delta_{d-1, J_2} 2^{d-1} w_{d-1}^{(2)} + \delta_{d, J_2} 2^{d-1} w_d^{(2)}.$$

(4.11)

Proof An argument similar to the proof of Theorem 4.9 shows the desired result. Let $d \geq 7$. In this case, by recalling Sect. 4.2.3, $\dim \operatorname{Harm}_i(\mathbb{R}^d)^{D_d} = 0$ for $i = 1, 2, 3, 5$. Moreover, $\operatorname{Harm}_4(\mathbb{R}^d)^{D_d}$ and $\operatorname{Harm}_6(\mathbb{R}^d)^{D_d}$ are spanned by f_4 and f_6, respectively. By substituting each corner vector v_k into f_4 and f_6, it holds that

$$f_4(v_k) = \begin{cases} \frac{1}{k}\left(1 - 3\frac{k-1}{d-1}\right), & \text{if } 1 \leq k \leq d-2, \\ -\frac{2}{d}, & \text{if } k = d-1, d, \end{cases}$$

$$f_6(v_k) = \begin{cases} \frac{1}{k^2}\left(1 - 15\frac{k-1}{d-1} + 30\frac{(k-1)(k-2)}{(d-1)(d-2)}\right), & \text{if } 1 \leq k \leq d-2, \\ \frac{16}{d^2}, & \text{if } k = d-1, d. \end{cases}$$

By Theorem 4.5, a pair $(\mathscr{X}(D_d, \{1, r\}, \{J_1, J_2\}), w)$ is a Euclidean 6-design supported by $\mathbb{S}^{d-1} \cup \mathbb{S}_r^{d-1}$ if and only if

$$0 = \sum_{x \in \mathscr{X}(D_d, \{1, r\})} w(x) \|x\|^{2j} f_i(x)$$

$$= \sum_{k \in J_1} \sum_{x \in v_k^{D_d}} w_k^{(1)} f_i(x) + \sum_{k \in J_2} \sum_{x \in (rv_k)^{D_d}} w_k^{(2)} r^{2j} f_i(x)$$

$$= \sum_{k \in J_1 \setminus \{d-1, d\}} 2^k \binom{d}{k} w_k^{(1)} f_i(v_k) + \delta_{d-1, J_1} 2^{d-1} w_{d-1}^{(1)} f_i(v_{d-1}) + \delta_{d, J_1} 2^{d-1} w_d^{(1)} f_i(v_d)$$

$$+ \sum_{k \in J_2 \setminus \{d-1, d\}} 2^k \binom{d}{k} w_k^{(2)} r^{i+2j} f_i(v_k) + \delta_{d-1, J_2} 2^{d-1} w_{d-1}^{(2)} r^{i+2j} f_i(v_{d-1})$$

$$+ \delta_{d, J_2} 2^{d-1} r^{i+2j} w_d^{(2)} f_i(v_d)$$

for $(i, j) = (4, 0), (4, 1), (6, 0)$, where the last equality follows from Lemma 4.1 and the fact that $|v_k^{D_d}| = 2^k \binom{d}{k}, k = 1, \ldots, d - 2$ and $|v_{d-1}^{D_d}| = |v_d^{D_d}| = 2^k$; see Sect. 4.2.3. Since $0 < r < 1$, the above equations for $(i, j) = (4, 0), (4, 1)$ are equivalent to (4.8). The above equation for $(i, j) = (6, 0)$ is equivalent to (4.9). Finally, (4.10) and (4.11) just follow from the definition of W_i, which completes the proof. $\square$

By Theorem 4.11, many families of D-optimal Euclidean 6-designs are provided. For positive integers k_1, k_2, let

$$G(k_1, k_2) := (d + 2 - 3k_1)(d + 2 - 3k_2) + 6(k_1 - 1)(k_2 - 1) + 2(d - 1).$$

Proposition 4.3 ([38], Proposition 1) Let $d \geq 8$.

(i) Let $J_1 = \{k, \ell, d\}$ and $J_2 = \{m, d\}$ with $k < \ell \leq d - 2$ and $m < \frac{d+2}{3}$. Assume that

$$m(-4 + 6\ell - d)(1 - W) + \ell(-4 + 6m - d)r^6 W > 0,$$
$$m(-4 + 6k - d)(1 - W) + k(-4 + 6m - d)r^6 W < 0,$$
$$2k\ell(-4 + 6m - d)(-1 + d)r^6 W + md(1 - W)G(k, \ell) > 0.$$

Then $(\mathscr{X}(D_d, \{1, r\}, \{J_1, J_2\}), w)$ with

$$w_k^{(1)} = \frac{k^2(d - 1)\{m(-4 + 6\ell - d)(1 - W) + \ell(-4 + 6m - d)r^6 W\} k!(d - k)!}{2^{k-1}(k - \ell)m(k - d)(d + 2)(d + 4)d!},$$

$$w_\ell^{(1)} = \frac{\ell^2(d - 1)\{m(-4 + 6k - d)(1 - W) + k(-4 + 6m - d)r^6 W\} \ell!(d - \ell)!}{2^{\ell-1}(\ell - k)m(\ell - d)(d + 2)(d + 4)d!},$$

$$w_d^{(1)} = \frac{d\left\{2k\ell(-4+6m-d)(-1+d)r^6W+md(1-W)G(k,\ell)\right\}}{2^{d-\ell}m(d+2)(d+4)(d-k)(d-\ell)},$$

$$w_m^{(2)} = \frac{m(d-1)Wm!(d-m)!}{2^{m-1}(d-m)(d+2)d!}, \quad w_d^{(2)} = \frac{(2-3m+d)dW}{2^{d-1}(d-m)(d+2)}$$

forms a D-optimal Euclidean 6-design.

(ii) *Let* $J_1 = \{k, \ell, d\}$ *and* $J_2 = \{j, m\}$ *such that* $k < \ell \le d-2$ *and* $j < m \le d-2$ *with* $j < \frac{d+2}{3}$ *and* $m > \frac{d+2}{3}$. *Assume that*

$$2jm(-4+6\ell-d)(-1+d)(1-W)+\ell dr^6WG(j,m) > 0,$$
$$2jm(-4+6k-d)(-1+d)(1-W)+kdr^6WG(j,m) < 0,$$
$$jmG(k,\ell)(1-W)+k\ell r^6WG(j,m) > 0.$$

Then $(\mathscr{X}(D_d, \{1, r\}, \{J_1, J_2\}), w)$ *with*

$$w_k^{(1)} = \frac{k^2k!(d-k)!\left\{2jm(-4+6\ell-d)(-1+d)(1-W)+\ell dr^6WG(j,m)\right\}}{2^k j(k-\ell)m(k-d)(d+2)(d+4)d!},$$

$$w_\ell^{(1)} = \frac{\ell^2\ell!(d-l)!\left\{2jm(-4+6k-d)(-1+d)(1-W)+kdr^6WG(j,m)\right\}}{2^\ell j(\ell-k)m(\ell-d)(d+2)(d+4)d!},$$

$$w_d^{(1)} = \frac{n^2\left\{jmG(k,\ell)(1-W)+k\ell r^6WG(j,m)\right\}}{2^{d-1}jm(d+2)(d+4)(d-k)(d-\ell)},$$

$$w_m^{(2)} = \frac{m(-2+3j-d)Wm!(-m+d)!}{2^m(j-m)(d+2)d!}, \quad w_j^{(2)} = \frac{j(-2+3m-d)Wj!(d-j)!}{2^j(m-j)(d+2)d!}$$

forms a D-optimal Euclidean 6-design.

Proof The result follows from (4.8), (4.9), (4.10), and (4.11). □

Example 4.7 Many D-optimal Euclidean 6-designs are produced by Proposition 4.3. For the case $(d, k, \ell, m) = (10, 1, 4, 2)$, combining Proposition 4.3 (i) and Table 3.1 in Sect. 3.3 yields that $(\mathscr{X}(D_{10}, \{1, 0.55259\}, \{\{1, 4, 10\}, \{2, 10\}\}), w)$ with

$$w_1^{(1)} = 0.00190373, \quad w_4^{(1)} = 0.000217705, \quad w_{10}^{(1)} = 0.000371822,$$
$$w_2^{(2)} = 0.0000834688, \quad w_{10}^{(2)} = 0.0000489075$$

forms a D-optimal Euclidean 6-design. Moreover, for the case $(d, k, \ell, m, j) = (10, 1, 4, 2, 6)$, by combining Proposition 4.3 (ii) and Table 3.1 in Sect. 3.3, $(\mathscr{X}(D_{10}, \{1, 0.55259\}, \{\{1, 4, 10\}, \{2, 6\}\}), w)$ with

$$w_1^{(1)} = 0.00190426, \quad w_4^{(1)} = 0.000217686, \quad w_{10}^{(1)} = 0.000371925,$$
$$w_2^{(2)} = 0.0000741945, \quad w_{10}^{(2)} = 2.31858 \times 10^{-6}$$

forms a D-optimal Euclidean 6-design. Note that these optimal values are rounded to no more than six significant figures for convenience.

Remark 4.6 In statistics language, $v_{d-1}^{D_d}$ or $v_d^{D_d}$ is the half fraction of a 2^d full factorial design. It can be also shown by Theorem 4.11 that $X = \{0\} \cup v_1^{D_d} \cup v_i^{D^d}$ ($i = d - 1, d$) cannot be of third-order rotatable; see also Theorem 5.1. Another statistical application of $v_i^{D_d}$ can be found in some recent work by Aoki et al. [1] in the area of computational algebraic statistics. In their work, the design matrix of a certain log-linear model for Box–Behnken $\{0\} \cup v_2^{D_d}$ designs is characterized in terms of *centrally symmetric configuration* associated with group D_d.

4.4   Combinatorial Thinning Methods

Victoir [42] proposes a *thinning method* for constructing cubature formulas with small number of points. The basic idea is to first prepare a B_d-invariant cubature formula in $\mathbb{R}^d$ and then reduce the number of points through combinatorial t-designs and orthogonal arrays (OAs). Hirao et al. [20] apply this thinning method to construct (optimal) Euclidean designs. While the corner vector method is powerful and produces a number of D-optimal Euclidean designs, designs obtained from that method have more experimental points than the Tchakaloff bound (3.10). The thinning method described in this section overcomes this inconvenience with combinatorial designs and OAs.

Let λ, t, d be positive integers with $t \leq d$, and let $\mathcal{K}$ be a set of distinct positive integers $k_1, \ldots, k_\ell \leq d$. A pair of a d-element set X and a collection $\mathcal{B}$ of subsets of X with sizes k_i in $\mathcal{K}$ is called a *t-wise balanced design*, say t-$(d, \mathcal{K}, \lambda)$, if

$$|\{B \in \mathcal{B} \mid T \subset B\}| = \lambda \quad \text{for every } T \in \binom{X}{t}.$$

In particular when $\mathcal{K} = \{k\}$, this is usually called a *(combinatorial) t-design* and denoted by t-(d, k, λ). Elements of X and $\mathcal{B}$ are called *points* and *blocks*, respectively. In statistics, 2- and 3-designs are called *balanced incomplete block* (BIB) *designs* and *doubly BIB designs*, respectively [21]. For a comprehensive treatment of combinatorial designs, we refer the reader to [24].

The number of blocks of a t-(d, k, λ) is given by

$$\lambda \frac{\binom{d}{t}}{\binom{k}{t}}.$$

Moreover, it is known (e.g., [24]) that for every $0 \leq t' \leq t$, a t-design is also a t'-design, namely, that the number of blocks containing a t'-element subset T' is given by

$$\lambda \frac{\binom{d-t'}{t-t'}}{\binom{k-t'}{t-t'}} = \lambda \frac{(d-t')(d-t'-1)\cdots(d-t+1)}{(k-t')(k-t'-1)\cdots(k-t+1)}.$$

In general, t-wise balanced designs do not have this property. A t-wise balanced design $(X, \mathscr{B})$ is said to be *regular* if for every $0 \le t' \le t$, the number of blocks containing a t'-element subset T' does not depend on the choice of T' [25]. When $t = 2$, some authors use the term "equireplicate" instead of "regular" [16].

Let s, ℓ, L, κ be nonnegative integers. An *orthogonal array (OA) of strength s, constraint ℓ and index κ* is an $L \times \ell$ matrix such that in every s columns, each of the 2^s ordered s-tuples of elements ± 1 appears exactly κ times among L rows; for example, see [17]. This is often denoted by $OA(L, \ell, 2, s)$. Parameter κ is not put in the notation because $\kappa = L/(2^s)$ by the definition of orthogonal array.

In the following result (Theorem 4.12), G (respectively, H) stands for a subgroup of B_d consisting of all permutations of the axes of $\mathbb{R}^d$ (respectively, all sign changes of the coordinates of a vector) which is isomorphic to symmetric group $\mathscr{S}_d$ (respectively, elementary abelian group $(\mathbb{Z}/2\mathbb{Z})^d$). Also, let $\tau(\omega)$ denote the nonzero components of $\omega \in \mathbb{R}^d$, and let

$$v_k^{(\alpha,\beta)} = \alpha \sum_{i=1}^{k} e_i + \beta \sum_{i=k+1}^{d} e_i.$$

The following Theorem 4.12 will directly follow from Propositions 5.1 and 5.2 (see Sect. 5.1).

Theorem 4.12 *(i) Let $\mathscr{K}$ be a subset of $\{1, 2, \ldots, d\}$. Let $X^{(\alpha,\beta)}$ be a subset of $\bigcup_{k \in \mathscr{K}} (v_k^{(\alpha,\beta)})^G$ such that $X^{(1,0)}$ is the set of characteristic functions (vectors) of a regular t-wise balanced design. Moreover, assume that there exists a G-invariant Euclidean $2e$-design of $\mathbb{R}^d$ of the form*

$$\mathscr{Q}[f] = c\left(\sum_{k \in \mathscr{K}} \frac{|X^{(\alpha,\beta)} \cap (v_k^{(\alpha,\beta)})^G|}{|X^{(\alpha,\beta)}|\binom{d}{k}} \sum_{x \in (v_k^{(\alpha,\beta)})^G} f(x) \right) + \sum_{x \in \bigcup_\ell v_\ell^G} w_\ell f(x)$$

with a positive integer c. Then

$$\mathscr{Q}[f] = \frac{c}{|X^{(\alpha,\beta)}|} \sum_{x \in X^{(\alpha,\beta)}} f(x) + \sum_{x \in \bigcup_\ell v_\ell^G} w_\ell f(x)$$

yields a Euclidean $2e$-design of $\mathbb{R}^d$.

(ii) For each $i = 1, \ldots, k$, let Y_i be the set of row vectors of $OA(|Y_i|, \tau(x_i), 2, 2e)$ corresponding to the runs. Assume that there exists an H-invariant Euclidean $2e$-design of $\mathbb{R}^d$ of the form

$$\mathscr{Q}[f] = \sum_{x \in \bigcup_{i=1}^{k} x_i^H} w_i f(x).$$

Then

$$\mathcal{Q}[f] = \sum_{x \in \bigcup_{i=1}^{k} x_i^H} \frac{2^{\tau(x_i)w_i}}{|Y_i|} f(x)$$

yields a Euclidean 2e-design of $\mathbb{R}^d$.

Remark 4.7 As far as the authors know, Theorem 4.12 (i) has been first formulated by Victoir [42, Theorem 3.3] only for combinatorial t-designs, and then extended by Nozaki and Sawa [34, Proposition 3.1] for regular t-wise balanced designs in general. See [30, 34] for the existence of regular t-wise balanced designs. The idea used in Theorem 4.12 (ii) goes back to a paper by Kôno [29] who utilizes particular types of orthogonal arrays to construct D-optimal Euclidean 4-designs on the hypercube. In [10], similar ideas are used to construct D-optimal designs of degree 3 on the hypercube; see also [35].

The following theorem is simple but theoretically very useful.

Theorem 4.13 ([20], Theorem 13) *Assume that there exists a B_d-invariant D-optimal Euclidean 2e-design. Then any design obtained by Theorem 4.12 preserves the D-optimality. Moreover, if there exists a B_d-invariant D-optimal Euclidean 2e-design with rational weights, then any design obtained by Theorem 4.12 also preserves the rationality of weights.*

Proof Note that the radii r_i and the total weights W_i of a D-optimal Euclidean design are not changed by Theorem 4.12.   □

Example 4.8 (i) According to Theorem 3.1, when $d = 7$, there exists a D-optimal invariant design ξ^* on $\mathbb{B}^7$ of degree 2 of type

$$\int_{\mathbb{B}^7} f(\omega)\,\xi^*(d\omega) = \frac{35}{36|\mathbb{S}^6|} \int_{\mathbb{S}^6} f(\omega)\,\rho(d\omega) + \frac{1}{36} f(0) \quad \text{for all } f \in \mathscr{P}_4(\mathbb{R}^7).$$

Then the following sum gives a B_7-invariant D-optimal Euclidean 4-design:

$$\frac{1}{36} \sum_{x \in (v_1^{(1,0)})^{B_7}} f(x) + \frac{1}{36} f(0).$$

Applying Theorem 4.12(i) to the 2-(7, 3, 1) design produces a D-optimal Euclidean 4-design of the form

$$\frac{1}{36} \sum_{x \in (v_1^{(1,0)})^{B_7}} f(x) + \frac{1}{36} f(0) = \frac{5}{36} \sum_{x \in X} f(x) + \frac{1}{36} f(0)$$

where X is a 56-point configuration induced by the 2-(7, 3, 1) design as

$$X = \{\tfrac{1}{\sqrt{3}}(\pm 1, \pm 1, 0, \pm 1, 0, 0, 0), \ \tfrac{1}{\sqrt{3}}(0, \pm 1, \pm 1, 0, \pm 1, 0, 0),$$
$$\tfrac{1}{\sqrt{3}}(0, 0, \pm 1, \pm 1, 0, \pm 1, 0), \ \tfrac{1}{\sqrt{3}}(0, 0, 0, \pm 1, \pm 1, 0, \pm 1),$$
$$\tfrac{1}{\sqrt{3}}(\pm 1, 0, 0, 0, \pm 1, \pm 1, 0), \ \tfrac{1}{\sqrt{3}}(0, \pm 1, 0, 0, 0, \pm 1, \pm 1),$$
$$\tfrac{1}{\sqrt{3}}(\pm 1, 0, \pm 1, 0, 0, 0, \pm 1)\};$$

see also Example 2.7. In analysis language, $X \cup \{(0, \dots, 0)\}$ has the same structure as the nodes of the Victoir minimal cubature formula for Gaussian integral [42, Sect. 5.1.1]. For a full generalization of the Victoir formula, see Hirao-Sawa [19, Corollary 4.5]. In design-theoretic language, this is the same arrangement as the Box–Behnken design No. 5 in [4, Table 4].
(ii) Applying Theorem 4.10 to the case where $d = 8$ with $J_1 = J_2 = \{2, 8\}$, a 736-point D-optimal Euclidean 6-design is obtained. Namely, this design improves the Tchakaloff bound (3.10) from $\binom{14}{6} = 3003$ points to 736 points; see also Proposition 3.2. This can be further reduced to a 480-point 6-design by use of an $OA(2^7, 8, 2, 6)$; see [17] for the existence of such OA.

Theorem 4.12 can be utilized to construct D-optimal Euclidean 6-designs with smaller number of points. As far as the authors know, the following Theorem 4.14 seems to be the first infinite family of D-optimal Euclidean 6-designs of $\mathbb{R}^d$ with $\mathcal{O}(d^5)$ points.

Theorem 4.14 ([20], Theorem 14) *Let $q = 3m + 1$ be an odd prime power with $m \notin \{1, 2, 4\}$. Then there exists a D-optimal Euclidean 6-design supported by $\mathbb{B}^{3m+2}$ with $O(m^5)$ points.*

Proof It is known [22] that if $(k - 1) \mid (q - 1)$, $k \nmid q$ and $k \notin \{3, 5\}$, then there exists a 3-$(q + 1, k + 1, k(k + 1)/2)$; see also [26, p. 82]. Let l be an integer with $2^{2\ell-2} \le m + 1 < 2^{2\ell}$, $\ell \ge 3$. Take an $OA(2^{6\ell-1}, 2^{2\ell}, 2, 7)$ with $\ell \ge 3$ which is the dual of Delsarte–Goethals code. The required result then follows from Theorems 4.10 and 4.12. The order of the number of points is computed as follows:

$$2^{6\ell+11} \cdot \frac{3(3m + 1)(3m + 2)}{2} = 2^{10} m^2 (3m)(3m + 1)(3m + 2) = O(m^5) \quad (m \to \infty).$$
$\square$

Remark 4.8 (i) The idea of thinning method (Theorem 4.12) can be found in some classical works in design of experiments, for example; see Box–Behnken designs [4]. Several subsequent works have been devoted to similar but more general constructions. For example, Das and Narasimham [9] discuss a generalization of Box–Behnken designs through combinatorial 3-designs (doubly BIB designs in their paper); see also Huda [21, Chap. 4] and Pesotchinsky [35]. The idea of Box–Behnken designs is to first prepare a BIB design and then extend it to an incomplete three-level factorial design. Whereas, our idea in Theorem 4.12 (i) is a reverse thinking. Namely, a B_d-orbit of corner vectors is first prepared so that an invariant theory is applicable to the construction of Euclidean designs and drastically reduce computational cost.

The idea of Theorem 4.12 (ii) at least goes back to Farrell et al. [10, pp. 136–137] (see also Kôno [29]), where a class of strength-four OA is employed to construct D-optimal Euclidean 4-designs with $O(d^3)$ observation points. Afterward, a large number of strength-four OA with fewer runs are found such as the Delsarte–Goethals OA; see e.g., [17, p. 103]. Note that these OA produce D-optimal Euclidean 4-designs with $O(d^2)$ observation points, which improves the result by Farrell et al. [10].

(ii) Theorem 4.14 uses an orthogonal array which is the dual of the Delsarte–Goethals code. The Delsarte–Goethals code belongs to a class of nonlinear codes and its construction is a bit complicated. Among many classical linear codes, a remarkably simple one is the BCH code; for example, see Hedayat et al. [17]. Cubature formulas reduced with the BCH code has somewhat larger sizes than cubature formulas reduced from the Delsarte–Goethals code, however, the former have a fairly simple structure and their constructions are quite simple. For more details about this, see Kuperberg [31].

4.5   Further Remarks and Open Questions

Euclidean designs of higher degree are not only of theoretical interest, as discussed in the previous section, but also useful in practice. In recent years, the construction theory for Euclidean designs of higher degree has gradually received attractive attention in the study of deterministic construction of (approximate) feature maps on machine learning (e.g., Dao et al. [8], Hirao and Sawa [18, Sect. 4]). However, as seen by Theorem 4.6, Euclidean designs of degree higher than or equal to 8 cannot be constructed just by use of our corner vector method for finite reflection groups A_d, B_d and D_d. In order to overcome this problem, Yamamoto et al. [43] propose the use of internally dividing points $v_{A,s}$ of corner vectors of group B_d as

$$v_{A,s} = \frac{1}{\sqrt{A+s}}\left(\sqrt{A}e_1 + \sum_{i=2}^{s+1} e_i\right), \quad A > 0, \ A \neq 1, \ 1 \leq s \leq d-1.$$

By substituting $v_{A,s}$ into f_4, f_6, $f_{8,1}$, $f_{8,2}$ defined in Sect. 4.2.2, it holds that

• for $d \geq 2$

$$f_4(v_{A,s}) = \frac{1}{(A+s)^2}\left\{A^2 + s - \frac{6}{d-1}\left(sA + \frac{s(s-1)}{2}\right)\right\},$$

- for $d \geq 3$

$$f_6(v_{A,s}) = \frac{1}{(A+s)^3}\left\{A^3 + s - \frac{15}{d-1}(As + A^2 s + s(s-1))\right.$$
$$\left. + \frac{180}{(d-1)(d-2)}\left(a^2\frac{s(s-1)}{2} + \frac{s(s-1)(s-2)}{3!}\right)\right\},$$

- for $d \geq 2$

$$f_{8,1}(v_{A,s}) = \frac{1}{(A+s)^4}\left\{A^4 + s - \frac{28}{d-1}(As + A^3 s + s(s-1))\right.$$
$$\left. + \frac{70}{d-1}\left(A^2 s + \frac{s(s-1)}{2}\right)\right\},$$

- for $d \geq 4$

$$f_{8,2}(v_{A,s}) = \frac{1}{(A+s)^4}\left\{A^2 s + \frac{(s-1)s}{2} - \frac{6}{d-2}\left(A(s-1)s + \frac{A^2(s-1)s}{2}\right.\right.$$
$$\left.\left. + \frac{(s-2)(s-1)s}{2}\right) + \frac{9s(-3+4A+s)(s-2)(s-1)}{2(d-3)(d-2)}\right\}.$$

Combining these equations and the Sobolev theorem for Euclidean designs (Theorem 4.3) provides B_d-invariant Euclidean 8-designs supported by three concentric spheres including the origin. For example, a B_6-invariant D-optimal Euclidean 8-design is given as follows:

$$\mathcal{Q}[f] = 0.000408226\sum_{x\in v^{B_6}_{5.98153,3}} f(x) + 0.00409238\sum_{x\in v^{B_6}_2} f(x)$$

$$+ 0.00116098\sum_{x\in v^{B_6}_4} f(x) + 0.00050098\sum_{x\in v^{B_6}_5} f(x)$$

$$+ 0.00119499\sum_{x\in(0.713334v_1)^{B_6}} f(x) + 0.000672181\sum_{x\in(0.713334v_3)^{B_6}} f(x)$$

$$+ 0.000336092\sum_{x\in(0.713334v_6)^{B_6}} f(x) + 0.0287936 f(0).$$

Note that these optimal values are rounded to no more than six significant figures for convenience.

Another approach for pushing up maximum degree is to relax the choice of (reflection) groups. For example, Sawa and Hirao [38] determine the maximum degree of the D-optimal Euclidean design constructed by corner vectors of exceptional groups H_3, H_4, F_4, E_6, E_7, and E_8. A statement in [38], without proof, is quoted here.

Theorem 4.15 *Let G be one of exceptional groups H_3, H_4, F_4, E_6, E_7 and E_8. Further let*

(i) $e \geq 6$ *if* $G = H_3$, (ii) $e \geq 12$ *if* $G = H_4$, (iii) $e \geq 6$ *if* $G = F_4$,

(iv) $e \geq 5$ *if* $G = E_6$, (v) $e \geq 6$ *if* $G = E_7$, (vi) $e \geq 8$ *if* $G = E_8$.

Then there is no choice of $\mathscr{R}$, $\mathscr{J}$, and w for which $(\mathscr{X}(G, \mathscr{R}, \mathscr{J}), w)$ is a D-optimal Euclidean 2e-design, independent of the number of supporting concentric spheres.

Sawa and Hirao [38] discover D-optimal Euclidean 10- and 12-designs for groups E_8 and H_4, respectively, and classifies maximum degree designs for groups H_3, H_4, and F_4 acting on the 3- and 4-dimensional Euclidean spaces.

An inspiring approach is to examine an appropriate finite subgroup of orthogonal group $\mathscr{O}(d)$ to find Euclidean designs of high degree. For example, it is interesting to find proper subgroups, e.g., cyclic subgroups (see Bose and Carter [2]), of groups A_d, B_d and D_d for which the maximum degree of invariant Euclidean designs becomes 8 or larger.

References

1. Aoki, S., Hibi, T., Ohsugi, H.: Markov-chain monte carlo methods for the box-behnken designs and centrally symmetric configurations. J. Stat. Theory Pract. **10**(1), 59–72 (2016)
2. Bose, R.C., Carter, R.L.: Complex representation in the construction of rotatable designs. Ann. Math. Stat. **30**, 771–780 (1959)
3. Bourbaki, N.: Lie Groups and Lie Algebras. Elements of Mathematics. Springer, Berlin (2002). Translated from the 1968 French original by Andrew Pressley
4. Box, G.E.P., Behnken, D.: Some new three level designs for the study of quantitative variables. Technometrics **2**(4), 455–475 (1960)
5. Box, G.E.P., Hunter, J.S.: Multi-factor experimental designs for exploring response surfaces. Ann. Math. Stat. **28**(1), 195–241 (1957)
6. Box, G.E.P., Hunter, J.S., Hunter, W.G.: Statistics for experimenters. In: Wiley Series in Probability and Statistics, 2nd edn. Wiley-Interscience, Wiley, Hoboken, NJ (2005)
7. Coxeter, H.S.M.: Introduction to Geometry. John Wiley & Sons Inc, New York-London-Sydney (1969)
8. Dao, T., De Sa, C., Ré, C.: Gaussian quadrature for kernel features. Adv. Neural. Inf. Process. Syst. **30**, 6109–6119 (2017)
9. Das, M.N., Narasimham, V.L.: Construction of rotatable designs through balanced incomplete block designs. Ann. Math. Stat. **33**(4), 1421–1439 (1962)
10. Farrell, R.H., Kiefer, J., Walbran, A.: Optimum multivariate designs. In: Proceedings of the Fifth Berkeley Symposium, vol. 1, pp. 113–138. University of California Press, Berkeley, California (1967)
11. Gaffke, N., Heiligers, B.: Algorithms for optimal design with application to multiple polynomial regression. Metrika **42**(3–4), 173–190 (1995)
12. Gaffke, N., Heiligers, B.: Computing optimal approximate invariant designs for cubic regression on multidimensional balls and cubes. J. Stat. Plan. Inference **47**(3), 347–376 (1995)
13. Gaffke, N., Heiligers, B.: Optimal and robust invariant designs for cubic multiple regression. Metrika **42**(1), 29–48 (1995)

14. Gaffke, N., Heiligers, B.: Minimum support invariant designs for multiple cubic regression. J. Stat. Plan. Inference **72**(1–2), 229–245 (1998)
15. Goethals, J.M., Seidel., J.J.: Cubature formulae, polytopes, and spherical designs. In: The Geometric Vein, pp. 203–218. Springer, Berlin (1981)
16. Gupta, S., Jones, B.: Equireplicate balanced block designs with unequal block sizes. Biometrika **70**(2), 433–440 (1983)
17. Hedayat, A.S., Sloane, N.J.A., Stufken, J.: Orthogonal Array. Springer Series in Statistics. Springer, Berlin (1999)
18. Hirao, M., Sawa, M.: On almost tight euclidean designs for rotationally symmetric integrals. Jpn. J. Stat. Data Sci. (to appear)
19. Hirao, M., Sawa, M.: On minimal cubature formulae of small degree for spherically symmetric integrals. SIAM J. Numer. Anal. **47**(4), 3195–3211 (2009)
20. Hirao, M., Sawa, M., Jimbo, M.: Constructions of Φ_p-optimal rotatable designs on the ball. Sankhya Ser. A **77**(1), 211–236 (2015)
21. Huda, S.: Rotatable designs: constructions and considerations in the robust design of experiments. Ph.D. thesis, Imperial College, University of London (1981)
22. Hughes, D.R.: On t-designs and groups. Am. J. Math. **87**, 761–778 (1965)
23. Humphreys, J.E.: Reflection Groups and Coxeter Groups. Cambridge University Press (1990)
24. Ionin, Y.J., van Trung, T.: Symmetric designs. In: Handbook of Combinatorial Designs, 2nd edn, pp. 110–124. CRC Press, Boca Raton, USA (2007)
25. Kageyama, S.: A property of t-wise balanced designs. Ars Combinatoria **31**, 237–238 (1991)
26. Khosrovshahi, G.B., Laue, R.: t-Designs with $t \geq 3$. In: Handbook of Combinatorial Designs, 2nd edn, pp. 79–101. CRC Press, Boca Raton, USA (2007)
27. Kiefer, J.: Optimum experimental designs V, with applications to systematic and rotatable designs. In: Proceedings of the 4th Berkeley Symposium, vol. 1, pp. 381–405. University California Press, Berkeley, California (1960)
28. Kiefer, J.: Optimum designs in regression problem II. Ann. Stat. **32**, 298–325 (1961)
29. Kôno, K.: Optimum design for quadratic regression on k-cube. Mem. Fac. Sci. Kyushu Univ. Ser. A **16**, 114–122 (1962)
30. Kramer, E., Kreher, D.: t-wise balanced designs. In: Handbook of Combinatorial Designs, 2nd edn, pp. 657–663. CRC Press, Boca Raton, USA (2007)
31. Kuperberg, G.: Numerical cubature using error-correcting codes. SIAM J. Numer. Anal. **44**(3), 897–907 (2006)
32. Njui, F., Patel, M.: Fifth order rotatability. Commun. Stat. Theory Methods **17**, 833–848 (1988)
33. Nozaki, H., Sawa, M.: Note on cubature formulae and designs obtained from group orbits. Can. J. Math. **64**(6), 1359–1377 (2012)
34. Nozaki, H., Sawa, M.: Remarks on Hilbert identities, isometric embeddings, and invariant cubature. Algebra i Analiz **25**(4), 139–181 (2013)
35. Pesotchinsky, L.: Φ_p-optimal second order designs for symmetric regions. J. Stat. Plan. Inference **2**(2), 173–188 (1978)
36. Pukelsheim, F.: Optimal design of experiments. In: Classics in Applied Mathematic, vol. 50. Society for Industrial and Applied Mathematics (SIAM), Philadelphia, PA (2006). Reprint of the 1993 original
37. Sali, A.: On the rigidity of spherical t-designs that are orbits of finite reflection groups. Des. Codes Cryptogr. **4**, 157–170 (1994)
38. Sawa, M., Hirao, M.: Characterizing D-optimal rotatable designs with finite reflection groups. Sankhya Ser. A **79**(1), 101–132 (2017)
39. Sawa, M., Xu, Y.: On positive cubature rules on the simplex and isometric embeddings. Math. Comp. **83**(287), 1251–1277 (2014)
40. Scheffé, H.: Experiments with mixtures. J. R. Stat. Soc. Ser. B **20**, 344–360 (1958)
41. Sobolev, S.L.: Cubature formulas on the sphere which are invariant under transformations of finite rotation groups (in Russian). Dokl. Akad. Nauk SSSR **146**, 310–313 (1962)
42. Victoir, N.: Asymmetric cubature formulae with few points in high dimension for symmetric measures. SIAM J. Numer. Anal. **42**(1), 209–227 (2004)
43. Yamamoto, H., Hirao, M., Sawa, M.: A construction of the fourth order rotatable designs invariant under the hyperoctahedral group. J. Stat. Plan. Inference **200**, 63–73 (2019)

Chapter 5
Euclidean Design Theory

The design of experiments, which has rapidly grown in connection with a wide range of fields such as engineering, psychology, etc., and which in particular has led to the development of the statistical quality control, is one of the most important areas in applied statistics. On the other hand, the theory of experimental design has been pursued mainly by quality engineers and statisticians, which is, however, still in its early stages from a viewpoint of mathematical description and elucidation. For example, in most standard textbooks on fractional factorial designs, the use of regular designs of two or three levels is emphasized and there are only a few descriptions concerning non-regular designs, with some exceptions such as Box–Behnken designs and Plackett–Burman designs. As far as the authors know, the only mathematical description dealing generally with multi-level multi-factorial designs is based on algebraic methods that can be used to express confounding relationship between factor effects.

In the present chapter, as a general mathematical description of experimental designs, we introduce the concept of generalized cubature formula and lay the foundation of *Euclidean Design Theory*. In the classical framework, a factorial design is an array with some alias relation, whereas in our new framework, a design is just a finite set of points in a given normed vector space. The present theory starts from such a simple geometric interpretation and then can be extended by incorporating some basic facts about kernel functions (reproducing kernels) introduced in Chap. 1.

Section 5.1 first gives the precise definition of generalized cubature formulas and translates binary orthogonal arrays and t-wise balanced designs in terms of cubature formulas. In Sect. 5.2 through 5.5, some fundamental results are selected in the classical cubature theory, as already seen in the previous chapters, and they are discussed what a natural generalization should be for them. Finally, Sect. 5.6 provides various further implications of some basic results in numerical analysis, algebraic combinatorics, and design of experiments that have already appeared in the previous chapters.

Some of the results given in this chapter can be also found in Sawa [29], where the description will be also updated below.

© The Author(s), under exclusive license to Springer Nature Singapore Pte Ltd. 2019   103
M. Sawa et al., *Euclidean Design Theory*, JSS Research Series in Statistics,
https://doi.org/10.1007/978-981-13-8075-4_5

5.1   Generalized Cubature Formula

Let μ be a (possibly infinite) positive measure on a set Ω. With the notation $\mathscr{L}^2(\Omega, \mu)$ as in Sect. 1.3, let $\mathscr{H}$ be a subspace of $\mathscr{L}^2(\Omega, \mu)$ embedded in $L^2(\Omega, \mu)$, the L^2-space on Ω. For convenience, let us call elements of $\mathscr{H}$ "L^2-integrable functions", and similarly for "L^1-integrable functions".

Definition 5.1 (*Generalized cubature formula*) Let A be a subset of Ω where a measure μ is concentrated, namely, $\mu(\Omega \setminus A) = 0$. A *generalized cubature formula* (*GCF*) *for* $\mathscr{H}$ is defined by points $x_1, \ldots, x_n \in A$ and weights $w_1, \ldots, w_n > 0$ such that

$$\int_\Omega f(\omega)\, \mu(d\omega) = \sum_{i=1}^n w_i f(x_i) \quad \text{for all } f \in \mathscr{H}.$$

In particular this is said to be of *Chebyshev-type* if w_i is independent of the choice of index i.

Example 5.1 A GCF for $\mathscr{P}_t(\Omega)$ is a cubature formula of degree t on a region $\Omega \subset \mathbb{R}^d$. In particular when $\Omega = \mathbb{S}^{d-1}$, a Chebyshev-type GCF for $\mathscr{P}_t(\mathbb{S}^{d-1})$ or a GCF for $\mathrm{Hom}_t(\mathbb{S}^{d-1})$ with respect to uniform measure $\rho/|\mathbb{S}^{d-1}|$ on unit sphere $\mathbb{S}^{d-1}$ is a spherical t-design or a Euclidean design of index t on $\mathbb{S}^{d-1}$, respectively. For the definition of index-type Euclidean design, see Sect. 2.4.

The following Proposition 5.1 gives an analytic definition of (regular) t-wise balanced designs in the context of cubature formulas, where a design is identified with the characteristic vectors of blocks; see Sect. 4.4 for the definition of t-wise balanced designs. It is emphasized that Proposition 5.1 and subsequent Proposition 5.2 play a crucial role in the construction of D-optimal Euclidean designs, as already seen in Theorems 4.3 and 4.14. For the significance of this point, the proof is given below.

Proposition 5.1 ([22]) *Let X be the characteristic functions (vectors) of the blocks of a given t-wise balanced design with a set V of d points and a family of blocks $\mathscr{B} = \bigcup_{i=1}^\ell \mathscr{B}_i$ of sizes $k_1, \ldots, k_\ell$, where each $\mathscr{B}_i$ consists of blocks of size k_i. Let*

$$\Omega := \bigcup_{i=1}^\ell \mathbb{S}^{d-1}_{\sqrt{k_i}}, \quad \mathbb{S}^{d-1}_{\sqrt{k_i}} := \{\omega \in \{0, 1\}^d \mid \|\omega\|_2^2 := \omega_1^2 + \cdots + \omega_d^2 = k_i\}$$

and

$$\mu := \sum_{i=1}^\ell \frac{|\mathscr{B}_i|}{|\mathscr{B}| \binom{d}{k_i}} \sum_{\omega \in \mathbb{S}^{d-1}_{\sqrt{k_i}}} \delta_\omega.$$

Then it holds that

$$\int_\Omega f(\omega)\, \mu(d\omega) = \frac{1}{|\mathscr{B}|} \sum_{x \in X} f(x) \quad \text{for all } f \in \mathscr{P}_t(\Omega). \tag{5.1}$$

Proof Clearly, for any $f \in \mathscr{P}_t(\Omega)$ and $a_1, \ldots, a_d \in \mathbb{N}$, it is seen that $f(\omega_1, \ldots, \omega_d)$ $= f(\omega_1^{a_1}, \ldots, \omega_d^{a_d})$ on $\Omega = \bigcup_{i=1}^\ell S_{\sqrt{k_i}}^{d-1}$. One may think of Ω as "binary spheres" and of μ as "uniform measure" on Ω. Hence, it suffices to show that (5.1) holds for monomials $f(\omega) = \prod_{i=1}^j \omega_i$, $1 \le j \le t$.

Let $0 \le t' \le t$. By the regularity, the number of blocks containing a t'-subset of V is a constant, say λ'. It turns out that

$$\sum_{i=1}^\ell |\mathscr{B}_i| \binom{k_i}{t'} = \lambda' \binom{d}{t'}. \tag{5.2}$$

To prove this, the size of the set $\mathscr{A} := \{(T', B) \in \binom{V}{t'} \times \mathscr{B} \mid T' \subset B\}$ is computed by double-counting technique. By using the characteristic function, say χ, it holds that

$$\sum_{i=1}^\ell |\mathscr{B}_i| \binom{k_i}{t'} = \sum_{i=1}^\ell \sum_{\substack{B \in \mathscr{B} \\ |B| = k_i}} \sum_{\substack{T' \subset B \\ T' \in \binom{V}{t'}}} \chi(T', B) = \sum_{B \in \mathscr{B}} \sum_{\substack{T' \subset B \\ T' \in \binom{V}{t'}}} \chi(T', B)$$

$$= |\mathscr{A}| = \sum_{T' \in \binom{V}{t'}} \sum_{T' \subset B \in \mathscr{B}} \chi(T', B) = \lambda' \binom{d}{t'}.$$

Now let $f(\omega) = \prod_{i=1}^{t'} \omega_i$. Then it follows from (5.2) and elementary calculations on binomial coefficients that

$$\frac{1}{|\mathscr{B}|} \sum_{x \in X} f(y) = \frac{1}{|\mathscr{B}|} \lambda'$$

$$= \frac{1}{|\mathscr{B}|} \sum_{i=1}^\ell |\mathscr{B}_i| \frac{\binom{k_i}{t'}}{\binom{d}{t'}}$$

$$= \sum_{i=1}^\ell \frac{|\mathscr{B}_i|}{|\mathscr{B}|} \cdot \frac{1}{\binom{d}{k_i}} \cdot \frac{\binom{k_i}{t'}\binom{d}{k_i}}{\binom{d}{t'}}$$

$$= \sum_{i=1}^\ell \frac{|\mathscr{B}_i|}{|\mathscr{B}|\binom{d}{k_i}} \cdot \binom{d - t'}{k_i - t'}$$

$$= \sum_{i=1}^\ell \frac{|\mathscr{B}_i|}{|\mathscr{B}|\binom{d}{k_i}} \sum_{\omega \in S_{\sqrt{k_i}}^{d-1}} f(\omega)$$

$$= \int_\Omega f(\omega)\, \mu(d\omega). \qquad \square$$

Another example will be looked at. As already defined in Sect. 4.4, a $k \times d$ array with entries ± 1, say $A := (a_{ij})$, is an orthogonal array (OA) of k runs, d constraints and strength t, if the number of columns containing a t-tuple T of ± 1 is independent of the choice of T. Our aim is again to regard array A as a GCF.

Proposition 5.2 ([33]) *Let X be the set of rows of a binary k × d orthogonal array with k runs, d constraints and strength t. Also let*

$$\Omega := \{\pm 1\}^d, \qquad \mu_{\pm} := \frac{1}{2^d} \sum_{\omega \in \Omega} \delta_{\omega}.$$

Then it holds that

$$\int_{\Omega} f(\omega)\, \mu_{\pm}(d\omega) = \frac{1}{|X|} \sum_{x \in X} f(x) \quad \text{for all } f \in \mathscr{P}_t(\Omega). \tag{5.3}$$

Proof As in the proof of Proposition 5.1, it suffices to show that (5.3) holds for monomials $f(\omega) = \prod_{i=1}^{j} \omega_i$ on Ω, where $0 \le j \le t$ with the convention $f \equiv 1$ if $j = 0$. One may think of Ω as "binary hypercube" and of μ as "uniform measure" on Ω. By the defining property of OA, it is easy to see that for $1 \le j \le t$, the both sides of (5.3) are zero. $\qquad\square$

Figure 5.1 may be helpful for readers to imagine a geometric realization of combinatorial t-designs and orthogonal arrays through Propositions 5.1 and 5.2. In this example, a BIB design with 3 treatments and 3 blocks of size 2 is first expressed as 3-dimensional points $(1, 1, 0)$, $(1, 0, 1)$, $(0, 1, 1)$, i.e., the midpoints of 3 edges of a regular simplex embedded in $\mathbb{R}^3$. An $OA(4, 2, 2, 2)$ is then regarded as four

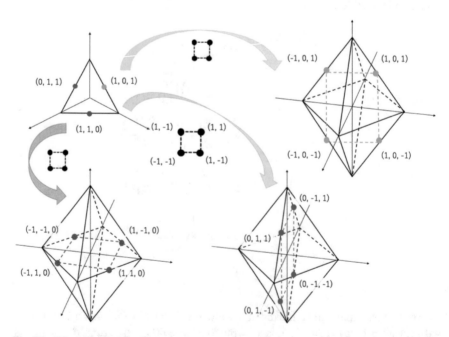

Fig. 5.1 What's "embedding"?

points $(1, 1)$, $(1, -1)$, $(-1, 1)$, $(-1, -1)$ of a square embedded in $\mathbb{R}^2$. Combining Propositions 5.1 and 5.2 then produces $3 \times 4 = 12$ points

$$
\begin{aligned}
&(1, 1, 0), \ (1, -1, 0), \ (-1, 1, 0), \ (-1, -1, 0), \\
&(1, 0, 1), \ (1, 0, -1), \ (-1, 0, 1), \ (-1, 0, -1), \\
&(0, 1, 1), \ (0, 1, -1), \ (0, -1, 1), \ (0, -1, -1)
\end{aligned}
$$

which forms a set of the midpoints (corner vectors) of 12 edges of a regular octahedron in $\mathbb{R}^3$. The first four points $(1, 1, 0)$, $(1, -1, 0)$, $(-1, 1, 0)$, $(-1, -1, 0)$ (colored red), which originally form a square in $\mathbb{R}^2$, are now embedded in the slicing plane perpendicular to the z-axis, and similarly for the remaining classes $\{(1, 0, 1), (1, 0, -1), (-1, 0, 1), (-1, 0, -1)\}$ (colored green) and $\{(0, 1, 1), (0, 1, -1), (0, -1, 1), (0, -1, -1)\}$ (colored blue).

The thinning method [15] for D-optimal Euclidean designs (Theorem 4.10) is based on Propositions 5.1 and 5.2, which produces an infinite series of optimal Euclidean 6-designs of $\mathbb{R}^d$ with $O(d^5)$ points (Theorem 4.14) that improves the upper bound by Farrell et al., [11] (see Proposition 3.2). This is an important application of Euclidean design theory to the optimal design theory.

Another important application of Euclidean design theory is able to evaluate the maximum order of classical rotatable designs. For example, let us take a look at Box–Behnken designs with n runs, based on a BIB design with d treatments and blocks of size k [4]. Such designs are just 2-dimensional $d \times n$ arrays in the classical framework of design of experiments, whereas they are, in our new framework, regarded as a finite set of points, say $X = \{x_1, \ldots, x_n\}$, in the Euclidean space $\mathbb{R}^d$. Now, with the multi-index notation $\omega^a = \omega_1^{a_1} \cdots \omega_d^{a_d}$ for $a = (a_1, \ldots, a_d) \in \mathbb{Z}_{\geq 0}^d$, consider the standard monomial regression of degree t as

$$
Y(\omega) = \sum_{\substack{a \in \mathbb{Z}_{\geq 0}^d \\ \|a\|_1 \leq t}} \alpha_a \omega^a + \varepsilon_\omega \tag{5.4}
$$

where $\mathbf{E}[\varepsilon_\omega] = 0$, $\text{Var}[\varepsilon_\omega] = \sigma^2$ for all $\omega \in \Omega$ and $\|a\|_1 = a_1 + \cdots + a_d$. The usual least square theory then shows (cf. [17, Eq. (4.1.3)]) that the normalized variance function of estimated response $\hat{Y}(\omega)$ at $\omega \in \Omega$ is given as

$$
d\left(\omega, \sum_{i=1}^n \delta_{x_i}\right) = (\omega^a)_{\|a\|_1 \leq t}^T \left(\sum_{i=1}^n x_i^a x_i^b\right)_{\|a\|_1, \|b\|_1 \leq t} (\omega^a)_{\|a\|_1 \leq t}, \tag{5.5}
$$

where $(\omega^a)_{\|a\|_1 \leq t}$ is the vector with all monomials ω^a as coordinates as in (3.5). Box and Hunter [5] prove that the following statements are equivalent:

(i) X is of eth order rotatable.[1]

---

[1] X is *of eth order rotatable* if (5.5) is a function of $\|\omega\|$; see also Remark 3.2.

(ii) For any integer $0 \leq \ell \leq 2e$ there exists some $c_{d,\ell} > 0$ such that for any $a \in \mathbb{Z}_{\geq 0}^d$ with $\|a\|_1 = \ell$,

$$\frac{1}{n}\sum_{i=1}^n x_i^a = \begin{cases} c_{d,\ell} \prod_{j=1}^d a_j!/(2^{\ell/2} \prod_{j=1}^d (a_j/2)!), & \text{if } a_j \text{ are even for all } j; \\ 0, & \text{otherwise.} \end{cases}$$

(5.6)

The following result evaluates the maximum value e for which Box–Behnken designs are of eth-order rotatable.

Theorem 5.1 *Let x_i be the set of blocks (characteristic vectors) of a BIB design with d treatments and blocks of size k. Then the following hold:*

(i) *Box–Behnken design $\{0\} \cup \bigcup_i x_i^{B_d}$ cannot be of third-order rotatable.*
(ii) *Box–Behnken design $\{0\} \cup \bigcup_i x_i^{B_d}$ is of second-order rotatable only if $d = 3k - 2$.*

Here B_d denotes the hyperoctahedral group defined in Sect. 4.1.

Proof As already seen in (2.6), conditions (5.6) are fulfilled for all moments up to degree $2e$ with respect to spherical integration $\int_{\mathbb{S}^{d-1}} \cdot \, d\rho/|\mathbb{S}^{d-1}|$. After appropriate scaling of points, X can be regarded as a Euclidean $2e$-design on $\mathbb{S}^{d-1}$. Statement (i) then immediately follows from Corollary 4.2 (i) in Sect. 4.3. Also, it is not difficult to prove (ii) by slightly modifying the argument concerning $f_4(v_i)$ in the proof of (4.4) in Theorem 4.9, i.e., by conventionally identifying inner sphere $\mathbb{S}_r^{d-1}$ with unit sphere $\mathbb{S}^{d-1}$ (and index set J_1 with J_2). $\qquad\square$

To regard Box–Behnken designs as B_d-invariant Euclidean designs is the key point of Theorem 5.1. Similar results may be also obtained for other classes of classical fractional factorial designs such as Plackett–Burman designs and central composite designs (e.g., [6]), which is, however, left for future work.

The subsequent sections discuss what a natural generalization should be for some fundamental results concerning classical polynomial-type cubature formulas, including the Tchakaloff theorem and the Sobolev theorem; recall Sects. 2.2 and 4.1. The Fisher inequality for BIB designs is also generalized in the framework of GCF and equality conditions are provided in terms of kernel functions (reproducing kernels).

5.2 Generalized Tchakaloff Theorem

The Tchakaloff theorem is a fundamental result in the classical theory of cubature formulas. In this section, as a starting point of the present theory, this theorem is generalized in the framework of GCF. It is here assumed that μ is a positive Borel measure, although most of the results given below can be proved even for positive measures in general.

Theorem 5.2 (Generalized Tchakaloff theorem) *Let μ be a (possibly infinite) positive Borel measure on a measurable space $(\Omega, \mathscr{F})$ concentrated in $A \in \mathscr{F}$. Also let $\mathscr{H}$ be a finite-dimensional vector space of L^1-functions on Ω. Then there exist a positive integer $n \leq \dim \mathscr{H} + 1$, points $x_1, \ldots, x_n \in A$, and weights $w_1, \ldots, w_n > 0$ such that*

$$\int_\Omega f(\omega)\, \mu(d\omega) = \sum_{i=1}^n w_i f(x_i) \quad \text{for all } f \in \mathscr{H}.$$

Moreover if $\mathscr{H}$ contains the constant function, then the evaluation of value n can be improved by one.

To prove this, some basic terminology and lemmas are reviewed.

Let μ be a measure on a measurable space $(\Omega, \mathscr{F})$ and $\phi : \Omega \to \mathbb{R}^N$ be a Borel measurable map. Denote by $\phi_* \mu$ the push-forward Borel measure on $\mathbb{R}^N$, following the notation of Bayer and Teichmann [3, p. 3036].

Lemma 5.1 ([3, Corollary 2]) *Let μ be a (possibly infinite) positive Borel measure on a measurable space $(\Omega, \mathscr{F})$ concentrated in $A \in \mathscr{F}$, and $\phi : \Omega \to \mathbb{R}^N$ be a Borel measurable map. Assume that the first moment of $\phi_* \mu$ exists, i.e.,*

$$\int_{\mathbb{R}^N} \|x\|\phi_*\, \mu(dx) < \infty.$$

Then there exist an integer $1 \leq n \leq N + 1$, points $x_1, \ldots, x_n \in A$, and weights $w_1, \ldots, w_n > 0$ such that

$$\int_\Omega \phi_j(\omega)\, \mu(d\omega) = \sum_{i=1}^n w_i \phi_j(\omega_i) \quad \text{for every } 1 \leq j \leq N + 1$$

where ϕ_j denotes the jth component of ϕ. Moreover, if ϕ_{N+1} is a constant, then the evaluation of the value n can be improved by one.

The following is a standard result in measure theory.

Lemma 5.2 *Let $(\Omega, \mathscr{F})$ be a measurable space and $\phi : \Omega \to \mathbb{R}^N$ be a map. Then ϕ is measurable if and only if for any $j = 1, \ldots, N$, the map $\phi_j : \Omega \to \mathbb{R}$ is measurable, where ϕ_j denotes the jth component of ϕ.*

Proof The "if" part is trivial. Conversely, assume that ϕ_j is measurable for any j. Then for any open subset $I_j \subset \mathbb{R}$, the inverse image $\phi_j^{-1}(I_j)$ is in $\mathscr{F}$. Since

$$\phi^{-1}(I_1 \times \cdots \times I_N) = \bigcap_{j=1}^N \phi_j^{-1}(I_j),$$

ϕ is measurable. $\square$

Proof of Theorem 5.2. Let $N = \dim \mathcal{H}$ and $f_1, \ldots, f_N$ be a basis of $\mathcal{H}$. Define

$$\phi : \Omega \to \mathbb{R}^N, \quad \omega \mapsto (f_1(\omega) \ldots, f_N(\omega)),$$

and consider the standard inner product on $\mathbb{R}^N$. Lemma 5.2 shows that the map ϕ is Borel measurable. It will be shown that the first moment of ϕ exists. In fact, note that

$$\|\phi(\omega)\| \leq |f_1(\omega)| + \cdots + |f_N(\omega)|$$

for any $\omega \in \Omega$. Since $f_1, \ldots, f_N$ are L^1,

$$\int_\Omega \|\phi(\omega)\| \, \mu(d\omega) \leq \sum_{i=1}^N \int_\Omega |f_i(\omega)| \, \mu(d\omega) < \infty.$$

The proof is thus completed by Lemma 5.1. $\square$

As a corollary of Theorem 5.2, a famous theorem due to Bayer and Teichmann [3] is obtained (Corollary 5.1), which generalizes the Tchakaloff theorem and also its various extensions and analogues as in [8, 25, 27].

Corollary 5.1 ([3]) *Let μ be a finite positive Borel measure on $\mathbb{R}^d$ concentrated in $A \in \mathcal{F}$, such that the moments up to degree t exist. Then there exist a positive integer $n \leq \dim \mathcal{P}_t(A)$, points $x_1, \ldots, x_n \in A$, and weights $w_1, \ldots, w_n > 0$ such that*

$$\int_{\mathbb{R}^d} f(\omega) \, \mu(d\omega) = \sum_{i=1}^n w_i f(x_i) \quad \text{for all } f \in \mathcal{P}_t(A).$$

Now, under some topological assumptions, a strengthening of Theorem 5.2 is proved.

Theorem 5.3 *Let Ω be a path-connected topological space and μ be a (possibly infinite) positive Borel measure on Ω with full support.[2] Denote by $\mathcal{H}$ a finite-dimensional vector space of continuous L^1-integrable functions on Ω. Then there exists a Chebyshev-type cubature for $\mathcal{H}$.*

The proof needs the following lemma which has been first proved by Seymour and Zaslavsky [30, Sect. 6].

Lemma 5.3 ([30]) *Let V be a finite-dimensional Euclidean space and $\tilde{\Omega}$ be a path-connected subset of V. Fix a vector v_0 in $\mathrm{Conv}(\tilde{\Omega})$. Then there exists a finite subset Y of $\tilde{\Omega}$ such that*

$$\frac{1}{|Y|} \sum_{y \in Y} y = v_0.$$

---

[2] Namely, $\mu(U) > 0$ for every non-void open set U.

Lemma 5.3 states that if v_0 belongs to $\mathrm{Conv}(\tilde{\Omega})$, then it can be represented as an average of points in some finite subset of $\tilde{\Omega}$.

Proof of Theorem 5.3. First take a basis $f_1, \ldots, f_n$ of $\mathcal{H}$ and consider the map $\phi : \Omega \to \mathbb{R}^n$ as in the proof of Theorem 5.2. Since any function in $\mathcal{H}$ is continuous, the map ϕ is also continuous. Put

$$v_0 := \left(\int_\Omega f_1(\omega)\, \mu(d\omega), \ldots, \int_\Omega f_n(\omega)\, \mu(d\omega) \right) \in \mathbb{R}^n.$$

To prove Theorem 5.3, it suffices to find a finite subset $Y \subset \Omega$ and a positive number λ such that

$$\lambda \sum_{y \in Y} \phi(y) = v_0.$$

By Theorem 5.2, there exists a cubature (X, w) for $\mathcal{H}$ on (Ω, μ). Put $\lambda_0 := \sum_{x \in X} w_x > 0$. Then

$$v_0 = \left(\int_\Omega f_1(\omega)\, \mu(d\omega), \ldots, \int_\Omega f_n(\omega)\, \mu(d\omega) \right)$$
$$= \left(\sum_{x \in X} w_x f_1(x), \ldots, \sum_{x \in X} w_x f_n(x) \right)$$
$$= \sum_{x \in X} w_x \phi(x)$$
$$= \sum_{x \in X} \frac{w_x}{\lambda_0} (\lambda_0 \phi(x)).$$

This means that $v_0 \in \mathrm{Conv}(\lambda_0 \phi(\Omega))$ with

$$\lambda_0 \phi(\Omega) := \{ \lambda_0 \phi(\omega) \mid \omega \in \Omega \} \subset \mathbb{R}^n.$$

Since map ϕ is continuous and Ω is path-connected, $\lambda_0 \phi(\Omega)$ is also path-connected. Hence, Lemma 5.3 shows the existence of a finite subset $\tilde{Y}$ of $\lambda_0 \phi(\Omega)$ such that

$$v_0 = \frac{1}{|\tilde{Y}|} \sum_{\tilde{y} \in \tilde{Y}} \tilde{y}.$$

For each $\tilde{y} \in \tilde{Y}$, there is $y \in \Omega$ such that $\tilde{y} = \lambda_0 \phi(y)$, which completes the proof. $\qquad\square$

The following result has been proved by the Seymour–Zaslavsky theorem [30] as a generalization of the mean value theorem for a univariate continuous function on a bounded interval.

Corollary 5.2 ([30]) *Let Ω be a path-connected topological space and μ be a finite positive Borel measure on Ω with full support. Further, let $\mathcal{H}$ be a* *finite-dimensional vector space of continuous L^1-integrable functions on Ω. Then there exists a Chebyshev-type cubature formula for $\mathscr{H}$.*

The Seymour–Zaslavsky theorem ensures the existence of Chebyshev-type cubature formulas on connected manifolds Ω, including existence theorems of Chebyshev-type cubature formulas for Grassmannian manifolds [1], compact symmetric spaces of rank one [2], flag manifolds [20], and the unitary groups [28]. Remark that all of such connected manifolds are also homogeneous groups of a Lie group. In general, if Ω is a homogeneous space of a Lie group G with finitely many connected components and μ is a G-invariant Haar measure, then connected components of Ω are isomorphic each other as measure spaces. In this case, the following Corollary 5.3 can also be proved.

Corollary 5.3 *Let Ω be a manifold with finitely many connected components and μ be a (possibly infinite) positive Borel measure on Ω with full support. Suppose that the connected components are isomorphic each other as measure spaces. Then for any finite-dimensional vector space $\mathscr{H}$ of continuous L^1-integrable functions on Ω, there exists a Chebyshev-type cubature formula for $\mathscr{H}$.*

Proof Denote by $(\Omega_1, \mu_1), \ldots, (\Omega_p, \mu_p)$ the connected components of (Ω, μ). Let $\mathscr{H}_i = \{ f|_{\Omega_i} \mid f \in \mathscr{H} \}$. Take a topological measure space (Ω_0, μ_0) isomorphic to any connected component of (Ω, μ). For each $i = 1, \ldots, p$, an isomorphism $\varphi_i : \Omega_0 \to \Omega_i$ is fixed between topological measure spaces. Denote by $\mathscr{H}_0$ the functional space on Ω_0 spanned by $\varphi_1^* \mathscr{H}_1, \ldots, \varphi_p^* \mathscr{H}_p$, where $\varphi_i^* \mathscr{H}_i$ is the pullback of $\mathscr{H}_i$ by φ_i for each i. Then $\mathscr{H}_0$ is also a finite-dimensional vector space of continuous L^1-integrable functions on Ω_0. Since Ω_0 is path-connected, by Theorem 5.3 there exists a Chebyshev-type cubature X_0 for $\mathscr{H}_0$ on (Ω_0, μ_0). Then, taking the disjoint union $X := \bigsqcup_{i=1}^p \varphi_i(X_0) \subset \Omega$, X is a Chebyshev-type cubature for $\mathscr{H}$ on (Ω, μ). $\qquad\square$

As an important implication of Corollary 5.3, a general existence theorem of optimal designs is obtained on a possibly disconnected experimental region Ω. The details will be coming up soon in Sect. 5.6.

5.3 Generalized Fisher Inequality

The previous section is devoted to general existence theorems of GCF with sufficiently large number of points. Now, how small can the number of points be? In this section, a lower bound is presented for the number of points needed and conditions for the equality are provided in terms of kernel functions (reproducing kernels).

Let μ be a positive measure on a set Ω. Further let $\mathscr{H}$ be a finite-dimensional Hilbert space over $\mathbb{K} = \mathbb{R}$ or $\mathbb{C}$ that consists of $\mathbb{K}$-valued L^2-integrable functions on Ω, with inner product as

$$(f, g)_{\mathscr{H}} = \int_{\Omega} f(\omega)\overline{g(\omega)}\, \mu(d\omega), \qquad f, g \in \mathscr{H}.$$

By the Riesz representation theorem (Theorem 1.4), for each $\omega \in \Omega$, there exists a function $f_\omega \in \mathcal{H}$ such that $f(\omega) = (f, f_\omega)_{\mathcal{H}}$ for any $f \in \mathcal{H}$. Consider a kernel function given by

$$K(\omega_1, \omega_2) := (f_{\omega_1}, f_{\omega_2})_{\mathcal{H}}.$$

It is remarked, by Proposition 1.3, that if $f_1, \ldots, f_N$ form an orthonormal basis of $\mathcal{H}$ with $N = \dim_{\mathbb{K}} \mathcal{H}$, then

$$K(\omega_1, \omega_2) = \sum_{j=1}^{N} f_j(\omega_1) \overline{f_j(\omega_2)} \quad \text{for all } \omega_1, \omega_2 \in \Omega$$

Note that functional space

$$\mathcal{H} \cdot \mathcal{H}^- := \mathrm{Span}_{\mathbb{K}} \{ f \cdot \overline{g} \mid f, g \in \mathcal{H} \}$$

is a finite-dimensional vector space of L^1-integrable functions on Ω.

Example 5.2 (Example 1.7, revisited). Let $\Omega = \mathbb{S}^{d-1}$ and ρ be the surface measure on $\mathbb{S}^{d-1}$. By (1.14), kernel polynomial $K_{\mathscr{P}}(\omega, \omega')$ for $\mathcal{H} := \mathscr{P}_e(\mathbb{S}^{d-1})$ with respect to $\int_{\mathbb{S}^{d-1}} \cdot \, d\rho / |\mathbb{S}^{d-1}|$ is expressed by

$$K_{\mathscr{P}}(\omega, \omega') = \sum_{i=0}^{e} Q_i(\langle \omega, \omega' \rangle), \qquad \omega, \omega' \in \mathbb{S}^{d-1}$$

where Q_i is the scaled Gegenbauer polynomial of degree i; see also [10, Definition 1.2]. Note that $\mathscr{P}_{2e}(\mathbb{S}^{d-1}) = \mathscr{P}_e(\mathbb{S}^{d-1}) \cdot \mathscr{P}_e(\mathbb{S}^{d-1})$.

The following is the main result of this section.

Theorem 5.4 (Generalized Fisher inequality) *Let X be a finite subset of Ω and w be a positive weight function. Then the following hold:*

(i) *Suppose that (X, w) is a cubature formula for $\mathcal{H} \cdot \mathcal{H}^-$. Then*

$$|X| \geq \dim_{\mathbb{K}} \mathcal{H}. \tag{5.7}$$

Moreover, the equality $|X| = \dim_{\mathbb{K}} \mathcal{H}$ holds if and only if

$$\sqrt{w_x w_y} \, K(x, y) = \delta_{xy} \quad \text{for every } x, y \in X. \tag{5.8}$$

(ii) *Suppose that (5.8) holds. Then $|X| \leq \dim_{\mathbb{K}} \mathcal{H}$. Moreover, the equality $|X| = \dim_{\mathbb{K}} \mathcal{H}$ holds if and only if (X, w) is a cubature formula for $\mathcal{H} \cdot \mathcal{H}^-$.*

To prove this, additional notations and lemmas are necessary.

Let $\mathbb{K}^X$ be the space of all $\mathbb{K}$-valued functions on X. Define a *scalar (inner) product* $(\cdot, \cdot)_X$ on $\mathbb{K}^X$ by

$$(\mathfrak{f}, \mathfrak{f}')_X := \sum_{x \in X} \mathfrak{f}(x) \overline{\mathfrak{f}'(x)} w_x \quad \text{for } \mathfrak{f}, \mathfrak{f}' \in \mathbb{K}^X.$$

Note that $\dim_{\mathbb{K}} \mathbb{K}^X = |X|$. Now consider the restriction map

$$\text{res} : \mathcal{H} \to \mathbb{K}^X, \quad f \mapsto f|_X.$$

Then it is obvious that res : $\mathcal{H} \to \mathbb{K}^X$ is a $\mathbb{K}$-linear map.

The following lemma not only plays a key role in the proof of Theorem 5.4 but also provides a geometric meaning of the definition of GCF.

Lemma 5.4 *The following are equivalent:*

(i) (X, w) is a cubature formula for $\mathcal{H} \cdot \mathcal{H}^-$.
(ii) The restriction map res : $\mathcal{H} \to \mathbb{K}^X$ *is a linear isometry.*

Also define the adjoint map of res : $\mathcal{H} \to \mathbb{K}^X$ by

$$\text{res}^* : \mathbb{K}^X \to \mathcal{H}.$$

Namely, for a function $\mathfrak{f}$ on X, res$^* \mathfrak{f}$ is the unique function in $\mathcal{H}$ such that

$$(f, \text{res}^* \mathfrak{f})_{\mathcal{H}} = (\text{res } f, \mathfrak{f})_X \quad \text{for all } f \in \mathcal{H}.$$

For each $x \in X$, let $\delta_x(y) = \delta_{xy}$ for every $y \in X$. Then $\{\delta_x \mid x \in X\}$ is a basis of $\mathbb{K}^X$. Note that res$^* \delta_x = w_x f_x$ for each $x \in X$, by the definition of res* and $f_x \in \mathcal{H}$.

Lemma 5.5 *The following are equivalent:*

(i) $\sqrt{w_x w_y} K(x, y) = \delta_{xy}$ for any $x, y \in X$.
(ii) The map res* : $\mathbb{K}^X \to \mathcal{H}$ *is a linear isometry.*

Proof Since $\{\delta_x \mid x \in X\}$ is a basis of $\mathbb{K}^X$, condition (ii) means that

$$(\text{res}^* \delta_x, \text{res}^* \delta_y)_{\mathcal{H}} = (\delta_x, \delta_y)_X \quad \text{for every } x, y \in X.$$

Note that $(\delta_x, \delta_y)_X = \sqrt{w_x w_y} \delta_{xy}$. Since res$^* \delta_x = w_x f_x$ for each $x \in X$, it holds that

$$(\text{res}^* \delta_x, \text{res}^* \delta_y)_{\mathcal{H}} = w_x w_y (f_x, f_y)_{\mathcal{H}} = w_x w_y K(x, y)$$

for every $x, y \in X$. $\square$

It is now ready to prove Theorem 5.4.

Proof of Theorem 5.4. (i) By Lemma 5.4 restriction map res : $\mathcal{H} \to \mathbb{K}^X$ is a linear isometry, which is injective and hence

$$\dim_{\mathbb{K}} \mathcal{H} \leq \dim_{\mathbb{K}} \mathbb{K}^X = |X|.$$

Note that $\dim_{\mathbb{K}} \mathcal{H} = |X|$ if and only if res : $\mathcal{H} \to \mathbb{K}^X$ is an isometric isomorphism, or equivalently, adjoint map res* : $\mathbb{K}^X \to \mathcal{H}$ is an isometric isomorphism. The result thus follows from Lemma 5.5.

(ii) By Lemma 5.5 adjoint map res* : $\mathbb{K}^X \to \mathcal{H}$ is a linear isometry, which is injective and hence

$$|X| = \dim_{\mathbb{K}} \mathbb{K}^X \leq \dim_{\mathbb{K}} \mathcal{H}.$$

The equality $\dim_{\mathbb{K}} \mathcal{H} = |X|$ holds if and only if res* : $\mathbb{K}^X \to \mathcal{H}$ is an isometric isomorphism. The required result thus follows from Lemma 5.4. □

By use of Theorem 5.4 (i), a famous inequality (Corollary 5.4) is presented for the number of blocks in a combinatorial $2e$-design. The first case to consider is when $e = 1$, i.e., the Fisher inequality for BIB designs [12]. The case of $e = 2$ was proved by Petrenjuk [23] and then generalized to all $2e$-designs by Ray-Chaudhuri and Wilson [26].

Corollary 5.4 (Wilson–Petrenjuk inequality) *Let $t = 2e$ and $d \geq k + e$. Assume that there exists a t-design with d points and b blocks of size k. Then it holds that*

$$b \geq \binom{d}{e}.$$

Proof As in Proposition 5.1, let $\Omega = S_{\sqrt{k}}^{d-1}$ and μ be the normalized counting measure. Letting $\mathcal{H} = \mathrm{Hom}_e(\Omega)$, it is remarked that

$$\mathcal{H} \cdot \mathcal{H}^- = \mathrm{Hom}_e(\Omega) \cdot \mathrm{Hom}_e(\Omega) = \mathrm{Hom}_t(\Omega).$$

Theorem 5.4 (i) shows that $|X| \geq \dim_{\mathbb{R}} \mathrm{Hom}_e(\Omega)$.

It remains to prove that

$$\dim_{\mathbb{R}} \mathrm{Hom}_e(\Omega) = \binom{d}{e}.$$

(This is trivial for $e = 1$, since all Dirac functions δ_x $(x \in X)$ give a basis of $\mathrm{Hom}_1(\Omega)$.) This is equivalent to showing that all monomials $\prod_{j=1}^{e} \omega_{i_j}$ of degree e are linearly independent over Ω. For this purpose, introduce a zero-one matrix of size $\binom{d}{i} \times \binom{d}{j}$, say M_{ij}, with rows indexed by the i-subsets of points and columns indexed by the j-subsets of points, and with entry 1 in row I_p and column J_q if $I_p \subset J_q$, where $0 \leq i \leq j \leq d$. Then the present goal is to show the positive definiteness of $M_{ek} M_{ek}^T$. In fact, note that the number of k-subsets, each of which contains a $(2e - i)$-subset I' but is disjoint from an i-subset I, is $\lambda \binom{d-2e}{k-2e+i} / \binom{d-2e}{k-2e}$, where λ

is the index of the design. By the definition of M_{ij}, it is seen that

$$M_{ek}M_{ek}^T = \sum_{i=0}^{e} \lambda \frac{\binom{d-2e}{k-2e+i}}{\binom{d-2e}{k-2e}} M_{ie}^T M_{ie} = \sum_{i=0}^{e-1} \lambda \frac{\binom{d-2e}{k-2e+i}}{\binom{d-2e}{k-2e}} M_{ie}^T M_{ie} + \lambda \frac{\binom{d-2e}{k-e}}{\binom{d-2e}{k-2e}} E\binom{d}{e}$$

which is clearly positive definite.   □

Theorem 5.4 (i) moreover generalizes various Fisher-type bounds that have been discussed separately in [1, 2, 13, 20, 28].

Now, how good is an estimate of lower bound (5.7)? Particularly for the case where $\mathscr{H} \cdot \mathscr{H}^- = \mathscr{P}_{2e}(\Omega)$, this has been a central theme in the theory of cubature formula in analysis, combinatorics and related areas. Several "tight" examples actually exist, for example, for BIB designs and sporadic examples of combinatorial 4- and 5-designs (e.g., [18]). Similar phenomena can be also observed for spherical designs and index-type Euclidean designs on the unit sphere, as already seen in Sect. 2.4. Also, Sect. 5.6 will describe one more class of tight GCF with respect to bound (5.7).

5.4 Generalized LP Bound

In this section linear programming (LP) bounds for spherical designs, which are substantially different from Fisher-type bounds (Theorem 5.4), are generalized in the framework of GCF. Throughout the present section let $\mathbb{K} = \mathbb{R}$ or $\mathbb{C}$.

Let (Ω, μ) be a measure space. Also let η be a L^1- and L^2-integrable function such that $\int_\Omega \eta(\omega)\,\mu(d\omega) \neq 0$. Define

$$\|\eta\|_2 := \left(\int_\Omega |\eta(\omega)|^2\,\mu(d\omega) \right)^{1/2} > 0.$$

Denote by K_η the kernel function (reproducing kernel) corresponding to 1-dimensional functional space $\mathbb{K}\langle\eta\rangle$. It is then easy to see that

$$K_\eta(\omega_1, \omega_2) = \frac{1}{\|\eta\|_2^2}\overline{\eta(\omega_1)}\eta(\omega_2) \quad \text{for all } (\omega_1, \omega_2) \in \Omega \times \Omega.$$

Let $\mathscr{H}$ be the space of all L^1- and L^2-integrable functions f such that $\int_\Omega f(\omega)\,\mu(d\omega) = 0$. Further, let $\{\mathscr{H}_i\}_{i \in I}$ be a family of subspaces of $\mathscr{H}$ with kernel functions $K_i(\cdot, \cdot)$, where I is an index set.

Now, assume that for an arbitrary set Υ there exists a surjection $\varpi : \Omega \times \Omega \to \Upsilon$. Let

$$\Delta(\Upsilon) := \{\varpi(\omega, \omega) \mid \omega \in \Omega\}.$$

Moreover assume that

(i) $\varpi^{-1}(\Delta(\Upsilon)) = \{(\omega, \omega) \mid \omega \in \Omega\}$;
(ii) $\eta(\omega)$ depends only on $\varpi(\omega, \omega)$, not depending on the choice of (ω, ω);
(iii) $K_\eta(\omega_1, \omega_2)$ depends only on $\varpi(\omega_1, \omega_2)$, not depending on the choice of (ω_1, ω_2);
(iv) $K_i(\omega_1, \omega_2)$ depends only on $\varpi(\omega_1, \omega_2)$ for each $i \in I$, not depending on the choice of (ω_1, ω_2).

Then the functions

$$\psi : \Delta(\Upsilon) \to \mathbb{K}, \qquad \psi(\varpi(\omega, \omega)) := \eta(\omega),$$
$$\tilde{Q}_\eta : \Upsilon \to \mathbb{K}, \qquad \tilde{Q}_\eta(\varpi(\omega_1, \omega_2)) := K_\eta(\omega_1, \omega_2),$$
$$\tilde{Q}_i : \Upsilon \to \mathbb{K}, \qquad \tilde{Q}_i(\varpi(\omega_1, \omega_2)) := K_i(\omega_1, \omega_2)$$

are well defined where $\omega \in \Omega$, $(\omega_1, \omega_2) \in \Omega \times \Omega$.

Consider formal $\mathbb{K}$-vector space $\mathbb{K}[\Upsilon]$ with basis Υ. Now use the notation $\xi \succeq 0$ if a vector $\xi = \sum_{\alpha \in \Upsilon} \xi_\alpha \alpha \in \mathbb{K}[\Upsilon]$ satisfies that $\xi_\alpha \geq 0$ for every $\alpha \in \Upsilon$.

Let T be a subset T of an index set I. Consider the convex subset of $\mathbb{K}[\Upsilon]$ given by

$$\Upsilon_{\eta, I, T} := \{ \xi \in \mathbb{K}[\Upsilon] \mid \xi \succeq 0, \ \tilde{Q}_\eta(\xi) = \frac{|\int_\Omega \eta(\omega)\, \mu(d\omega)|^2}{\|\eta\|_2^2}, \ \tilde{Q}_t(\xi) = 0, \ \tilde{Q}_i(\xi) \geq 0$$
$$\text{for every } t \in T \text{ and } i \in I \setminus T \},$$

where $\tilde{Q}_i$ is defined by $\sum_{i \in \Upsilon} \xi_i Q(i)$ for $\xi = \sum_{i \in \Upsilon} \xi_i i$ ($\xi_i \in \mathbb{K}$) and similarly for $\tilde{Q}_\eta$. Define

$$\mathscr{H}_{\eta, T} := \mathbb{K}\langle \eta \rangle \oplus \mathrm{Span}_\mathbb{K}\{ f \mid f \in \mathscr{H}_t, \ t \in T \}.$$

Then the following bound can be presented.

Theorem 5.5 (Generalized LP bound) *Let (X, w) be a cubature formula for $\mathscr{H}_{\eta, T}$ and*

$$\psi(\xi) := \sum_{\alpha \in \Delta(\Upsilon)} \xi_\alpha \psi(\alpha).$$

Then an inequality

$$|X| \geq \frac{|\int_\Omega \eta(\omega)\, \mu(d\omega)|^2}{\sup\{\psi(\xi) \mid \xi \in \Upsilon_{\eta, I, T}\}}$$

holds. Moreover, the equality holds only if $|\eta(x)|w_x$ is constant on X.

Proof It follows from the Cauchy–Schwarz inequality that

$$0 < \left| \int_\Omega \eta(\omega)\, \mu(d\omega) \right|^2 = \left| \sum_{x \in X} \eta(x) w_x \right|^2 \leq |X| \sum_{x \in X} |\eta(x)|^2 w_x^2$$

and hence

$$|X| \geq \frac{|\int_{\Omega} \eta(\omega)\, \mu(d\omega)|^2}{\sum_{x \in X} |\eta(x)|^2 w_x^2}.$$

It remains to show that

$$\sum_{x \in X} |\eta(x)|^2 w_x^2 \leq \sup\{\psi(\xi) \mid \xi \in \Upsilon_{\eta,I,T}\}. \tag{5.9}$$

For $\xi_{X,w} \in \mathbb{K}[\Upsilon]$, define the αth entry ($\alpha \in \Upsilon$) of $\xi_{X,w}$ as follows:

$$(\xi_{X,w})_\alpha := \sum_{\substack{(x,y) \in X \times X \\ \varpi(x,y)=\alpha}} w_x w_y.$$

Then it holds that $\sum_{x \in X} |\eta(x)|^2 w_x^2 = \psi(\xi_{X,w})$. Since $\xi_{X,w} \in \Upsilon_{\eta,I,T}$, (5.9) follows. It is clear that a necessary condition for equality is the constancy of $|\eta(x)|w_x$ for all $x \in X$. $\qquad\qquad\square$

Example 5.3 Let $\Omega = \mathbb{S}^{d-1}$ and μ be the uniform measure on $\mathbb{S}^{d-1}$. With the univariate polynomial R_e defined in (1.15), let $\eta(\omega) = R_e(\langle \omega, \omega \rangle)$. Then $\eta(\omega) \equiv R_e(1)$ and the corresponding kernel $K_\eta(\omega_1, \omega_2) \equiv 1$ on $\mathbb{S}^{d-1}$. Taking $\mathcal{H} = \mathscr{P}_{2e}(\mathbb{S}^{d-1})$ and $\mathcal{H}_i = \mathrm{Harm}_i(\mathbb{S}^{d-1})$ provides the Fischer decomposition $\mathcal{H} = \bigoplus_{i=0}^{2e} \mathcal{H}_i$ as in (1.13). By the addition formula (Theorem 1.5), the kernel K_i corresponding to $\mathcal{H}_i$ is given as

$$K_i(\omega_1, \omega_2) = \sum_{j=1}^{h_i^d} \phi_{i,j}(\omega_1)\phi_{i,j}(\omega_2) = Q_i(\langle \omega_1, \omega_2 \rangle).$$

Regard ϖ as Euclidean inner product $\langle \cdot, \cdot \rangle$. All conditions (i) through (iv) are satisfied for $\varpi, \eta, K_\eta, K_i$ given above. Of course, in this case, functions ψ and $\tilde{Q}_\eta$ are constant functions. Let $I = \{0, 1, \ldots, 2e\}$ and $T = I \setminus \{0\}$. Condition $\tilde{Q}_\eta(\xi) = |\int_{\Omega} \eta(\omega)\, \mu(d\omega)|^2 / \|\eta\|_2^2$ concerning $\Upsilon_{\eta,I,T}$ is automatically satisfied. Condition $\tilde{Q}_t(\xi) = 0$ implies that for $1 \leq t \leq 2e$, a finite weight summation of $\tilde{Q}_t$ at ξ is equal to zero; for example, see Sect. 1.4. Since $\mathbb{K}\langle \eta \rangle$ consists of the constant functions,

$$\mathcal{H}_{\eta,T} = \mathbb{K}\langle \eta \rangle \oplus \mathrm{Span}_{\mathbb{R}}\{f \mid f \in \mathcal{H}_t,\ t \in T\}$$

$$= \mathrm{Harm}_0(\mathbb{S}^{d-1}) \oplus \bigoplus_{i=1}^{2e} \mathrm{Harm}_i(\mathbb{S}^{d-1}) = \bigoplus_{i=0}^{2e} \mathrm{Harm}_i(\mathbb{S}^{d-1}) = \mathscr{P}_{2e}(\mathbb{S}^{d-1}).$$

Delsarte et al. [10] choose a good candidate for $\sup\{\psi(\xi) \mid \xi \in \Upsilon_{\eta,I,T}\}$ and thereby obtains the LP bound for cubature formulas with respect to the uniform measure on $\mathbb{S}^{d-1}$. A detailed explanation on their work can be also found in Sect. 2.4 of the present book.

Remark 5.1 To get nontrivial bounds using Theorem 5.5, it is necessary to consider how to evaluate $\sup\{\psi(\xi) \mid \xi \in \Upsilon_{\eta,I,T}\}$. This is a kind of linear programming problems. It is quite likely that Delsarte et al. [10] have first treated LP bounds for spherical designs which are much better than Fisher-type bounds. Another good LP bound has been found by Xu [34] for cubature formulas with respect to the multivariate Jacobi weight function on unit ball $\mathbb{B}^d$.

5.5 Generalized Sobolev Theorem

As already seen in Chap. 4, the classical invariant theory can be naturally absorbed in the theory of polynomial-type cubature formulas, producing a huge number of Euclidean designs and cubature formulas. In this section, the Sobolev theorem (Theorem 4.2) is generalized in the framework of GCF.

Let G be a finite subgroup of orthogonal group $\mathcal{O}(d)$. For a measure space (Ω, μ) in $\mathbb{R}^d$, (Ω, μ) is said to be *G-invariant* if Ω and μ are invariant under G, respectively. This terminology is also used for weighted pair (X, w), where X is a finite subset of Ω and w is a positive weight function on X. Given an $\mathbb{R}$-function f on Ω, consider the action of $g \in G$ on f as follows:

$$f^g(\omega) = f(\omega^{g^{-1}}), \quad \omega \in \Omega.$$

Fix $\mathbb{R}$-functions $f_1, \ldots, f_\ell$ on Ω and define

$$\mathcal{H} = \mathrm{Span}_{\mathbb{R}}\{f_i^g \mid 1 \leq i \leq \ell, \ g \in G\}.$$

A function f is said to be *G-invariant* if $f^g = f$ for every $g \in G$. Denote by $\mathcal{H}^G$ the set of G-invariant functions in $\mathcal{H}$. Define *Reynolds operator R* as follows (e.g., [7]):

$$R : \mathcal{H} \to \mathcal{H}, \quad R(f) = \frac{1}{|G|} \sum_{g \in G} f^g.$$

Lemma 5.6 *The following hold:*

(i) R is a linear map.
(ii) $R(f) \in \mathcal{H}^G$ for any $f \in \mathcal{H}$. In particular $R(f) = f$ for any $f \in \mathcal{H}^G$.

Proof (i) is straightforward. To prove (ii), fix $h \in G$. Since G is invariant under multiplication by h, it turns out that

$$R(f)^h(\omega) = \frac{1}{|G|} \sum_{g \in G} f^h(\omega^{g^{-1}}) = \frac{1}{|G|} \sum_{g \in G} f(\omega^{g^{-1}h^{-1}})$$

$$= \frac{1}{|G|} \sum_{g \in G} f^{hg}(\omega) = \frac{1}{|G|} \sum_{g \in G} f^g(\omega) = R(f)(\omega)$$

which shows $R(f) \in \mathcal{H}^G$. Proof of the latter part is left as an exercise. $\square$

Theorem 5.6 (Generalized Sobolev theorem) *Let (Ω, μ) be a G-invariant measure space and (X, w) be a G-invariant pair. Then the following are equivalent:*

(i) (X, w) is a GCF for $\mathcal{H}$.
(ii) (X, w) is a GCF for $\mathcal{H}^G$.

Proof The implication "(i) to (ii)" is trivial. Next assume (ii). Let $f \in \mathcal{H}$. Define

$$\phi := R(f) \left(= \frac{1}{|G|} \sum_{g \in G} f^g \right)$$

which is G-invariant by Lemma 5.6 (ii). By the G-invariance of (Ω, μ), it holds that

$$\int_{\Omega} f(\omega) \, \mu(d\omega) = \frac{1}{|G|} \sum_{g \in G} \int_{\omega \in \Omega} f(\omega) \, \mu(d\omega)$$

$$= \int_{\Omega} \frac{1}{|G|} \sum_{g \in G} f(\omega) \, \mu(d\omega)$$

$$= \int_{\Omega^g} \frac{1}{|G|} \sum_{g \in G} f(\omega^{g^{-1}}) \, \mu(d\omega^{g^{-1}})$$

$$= \int_{\Omega} \frac{1}{|G|} \sum_{g \in G} f^g(\omega) \, \mu(d\omega)$$

$$= \int_{\Omega} \phi(\omega) \, \mu(d\omega).$$

Similarly from the G-invariance of (X, w), define $X = \bigcup_i x_i^G$ and $w_i = w(x)$ for $x \in x_i^G$. Then it follows that

$$\int_{\Omega} f(\omega) \, \mu(d\omega) = \int_{\Omega} \phi(\omega) \, \mu(d\omega)$$

$$= \sum_i w_i \sum_{x \in x_i^G} \frac{1}{|G|} \sum_{g \in G} f^g(x)$$

$$= \sum_i w_i \frac{1}{|G|} \sum_{g \in G} \sum_{x \in x_i^G} f^g(x),$$

$$= \sum_i w_i \frac{1}{|G|} \sum_{g \in G} \sum_{x \in x_i^G} f(x),$$

$$= \sum_i w_i \sum_{x \in x_i^G} f(x),$$

$$= \sum_{x \in X} w_x f(x),$$

which completes the proof of the theorem. □

When $\Omega = \mathbb{S}^{d-1}$ and $\rho/|\mathbb{S}^{d-1}|$ is the uniform measure on $\mathbb{S}^{d-1}$, Theorem 5.6 just coincides with the Sobolev theorem (Theorem 4.2). The following provides a simple upper bound for the number of generators of $\mathcal{H}^G$.

Proposition 5.3 $\mathcal{H}^G = \text{Span}_{\mathbb{R}}\{R(f_i) \mid 1 \leq i \leq \ell\}$. *In particular, the number of generators of $\mathcal{H}^G$ is bounded above by ℓ.*

Proof Since $R(f_i^g) = R(f_i)$ for any $g \in G$, the result follows from Lemma 5.6 (i). □

As for the polynomial case (See Sect. 4.2), there is the possibility that the number of generators of $\mathcal{H}^G$ could be much smaller than ℓ, depending on the choices of G and f_i.

Theorem 5.6 drastically reduces the computational cost of finding cubature formulas for a given functional space. In the next section Theorem 5.6 is utilized to obtain a new class of cubature formulas for functional space $\mathcal{H} \cdot \mathcal{H}^-$ with $\mathcal{H} = \bigoplus_{i=e-1}^{e} \text{Hom}_i(\mathbb{R}^2)$, which are tight with respect to the generalized Fisher inequality (5.7).

5.6 Conclusion and Further Implications

In the present chapter, the notion of generalized cubature formula has been introduced as a unified treatment of various mathematical and/or statistical objects, and such basic results as the Tchakaloff theorem, the Sobolev theorem, LP bounds and the Fisher inequality are extended. As mentioned several times in the previous section, a theoretical impact of the GCF framework will be the "thinning construction" of optimal Euclidean designs with small number of experimental points. It is also emphasized that Euclidean design theory finds meaningful application in the evaluation of the maximum value t for which classical fractional factorial designs such as Box–Behnken designs and central composite designs are of tth order rotatable (Theorem 5.1). Nevertheless, the present theory is just in the starting stage and far from completion. For example, the Möller bound (Theorem 2.3) is, as already checked in Chap. 2, an important result in the classical theory of cubature formulas, which has not been discussed here in the framework of GCF. Hopefully, this direction of research will make further progress with the effect that practical and/or theoretical applications follow.

The final section is now closed with further implications of the basic results in connection with some selected topics in analysis, algebra, and design of experiments.

As explained in Sect. 5.3, there have been various examples of tight combinatorial designs, tight spherical designs, and tight index-type spherical designs. One more new class of tight GCF with respect to bound (5.7) is here presented.

Theorem 5.7 *Let $\Omega = \mathbb{R}^2$ and $\mu(d\omega) = W(\|\omega\|)\, d\omega$. For a positive integer e, let $\mathcal{H} = \bigoplus_{i=e-1}^{e} \mathrm{Hom}_i(\Omega)$ and $\alpha_e = \int_{\Omega} \|\omega\|^{2e} \mu(d\omega)$. For $0 \le i \le 2e$ define*

$$x_i := \sqrt{\frac{\alpha_e}{\alpha_{e-1}}} \cdot \left(\cos\left(\frac{2\pi i}{2e+1}\right),\ \sin\left(\frac{2\pi i}{2e+1}\right) \right), \qquad w_i := \frac{1}{2e+1} \cdot \frac{\alpha_{e-1}^e}{\alpha_e^{e-1}}.$$

Then

$$\int_{\Omega} f(\omega)\, \mu(d\omega) = \sum_{i=0}^{2e} w_i f(x_i) \quad \text{for all } f \in \mathcal{H} \cdot \mathcal{H}^-. \tag{5.10}$$

Moreover, this configuration is tight with respect to bound (5.7).

Proof Denote by X the vertex set of the regular $(2e+1)$-gon inscribed in the circle of radius $\sqrt{\alpha_e/\alpha_{e-1}}$. By Theorem 5.6 it suffices to show (5.10) for $f \in (\mathcal{H} \cdot \mathcal{H}^-)^{I_2(2e+1)}$, where $I_2(n)$ denotes the Dihedral group of order $2n$ acting on a regular n-gon in $\mathbb{R}^2$; see Sect. 4.1. Since

$$\mathcal{H} \cdot \mathcal{H}^- = \bigoplus_{i=e-1}^{e} \mathrm{Hom}_i(\Omega) \cdot \bigoplus_{i=e-1}^{e} \mathrm{Hom}_i(\Omega) = \bigoplus_{i=2e-2}^{2e} \mathrm{Hom}_i(\Omega),$$

the space $(\mathcal{H} \cdot \mathcal{H}^-)^{I_2(2e+1)}$ is spanned by $\|\omega\|_2^{2e-2}$ and $\|\omega\|_2^{2e}$. By the definition of w_i and x_i, it holds that

$$\sum_{i=0}^{2e} w_i \|x_i\|_2^{2e-2} = \sum_{i=0}^{2e} \frac{1}{2e+1} \cdot \frac{\alpha_{e-1}^e}{\alpha_e^{e-1}} \cdot \left(\sqrt{\frac{\alpha_e}{\alpha_{e-1}}} \right)^{2e-2} = \alpha_{e-1} = \int_{\Omega} \|\omega\|_2^{2e-2}\, \mu(d\omega),$$

$$\sum_{i=0}^{2e} w_i \|x_i\|_2^{2e} = \sum_{i=0}^{2e} \frac{1}{2e+1} \cdot \frac{\alpha_{e-1}^e}{\alpha_e^{e-1}} \cdot \left(\sqrt{\frac{\alpha_e}{\alpha_{e-1}}} \right)^{2e} = \alpha_e = \int_{\Omega} \|\omega\|_2^{2e}\, \mu(d\omega),$$

which prove the first assertion. The latter assertion is shown by noting that $|X| = 2e+1 = \dim \bigoplus_{i=e-1}^{e} \mathrm{Hom}_i(\Omega)$. □

Remark 5.2 The above examples of tight GCF are closely related to a class of integration formulas called *Laplacian-type cubature formulas* [14], which are closely connected to the so-called *Hermite interpolation* in numerical analysis and *Gauss mean value property* in harmonic analysis. For more details, see [14] and [31].

The left-hand side of Fig. 5.2 illustrates a tight cubature formula for

$$(\mathrm{Hom}_2(\mathbb{B}^2) \oplus \mathrm{Hom}_1(\mathbb{B}^2))^2 = \mathrm{Hom}_4(\mathbb{B}^2) \oplus \mathrm{Hom}_3(\mathbb{B}^2) \oplus \mathrm{Hom}_2(\mathbb{B}^2)$$

where the origin is not crucial as cubature points. The right-hand side of Fig. 5.2 (see also Sect. 2.2) describes a tight cubature formula for $\mathscr{P}_4(\mathbb{B}^2)$, where the origin plays an important role because any translate of polynomials of degree at most 4 is taken

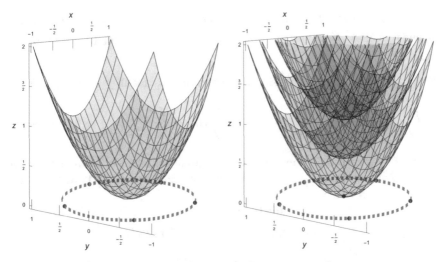

Fig. 5.2 Tight cubature for $(\mathrm{Hom}_2(\mathbb{B}^2) \oplus \mathrm{Hom}_1(\mathbb{B}^2))^2$ (left) and $\mathscr{P}_4(\mathbb{B}^2)$ (right)

into consideration. As already seen in the previous two chapters, a natural statistical model for the left is the standard quadratic regression (5.4), whereas a model for the right may be proposed by deleting the constant term from (5.4) where only the first- and second-order terms are of interest.

To find other integrals and functional spaces for which the generalized Fisher-type bound is sharp will be a challenging and interesting problem.

As another implication of the basic results, we give an alternative proof of a famous theorem by Reznick [27, Theorem 9.5] on Hilbert identities.

Theorem 5.8 ([27]) *A binary Hilbert identity with $e + 1$ 2eth powers is uniquely given by*

$$(x_1^2 + x_2^2)^e = \sum_{i=1}^{e+1} (a_i x_1 + b_i x_2)^{2e}.$$

Here the $2e + 2$ points $\{\pm(a_i, b_i) \mid i = 1, \ldots, e + 1\}$ are the vertices of a regular $(2e + 2)$-gon inscribed in the circle of radius $2^{-1}(\binom{2e}{e}/(e + 1))^{1/2e}$ about the origin.

By Proposition 2.3, this is equivalent to the statement that a GCF for $\mathrm{Hom}_{2e}(\mathbb{R}^2)$ with $e + 1$ points is uniquely determined by

$$\frac{1}{|\mathbb{S}^1|} \int_{\mathbb{S}^1} f(\omega)\, \rho(d\omega) = \frac{1}{e+1} \sum_{i=1}^{e+1} f\left(\cos\left(\frac{i\pi}{e+1} + \theta \right), \sin\left(\frac{i\pi}{e+1} + \theta \right) \right), \quad \theta \in [0, 2\pi).$$

Lemma 5.7 *If points $x_1, \ldots, x_n$ and weights $w_1, \ldots, w_n$ form a GCF for $\mathrm{Hom}_{2e}(\mathbb{R}^2)$ with respect to $\int_{\mathbb{B}^2} \cdot\, d\omega$, then points $x_i / \|x_i\|_2$ and weights $\|x_i\|_2^{2e} w_i / \int_0^1 r^{2e+1}\, dr$ form a GCF for $\mathrm{Hom}_{2e}(\mathbb{R}^2)$ with respect to $\int_{\mathbb{S}^1} \cdot\, \rho(d\omega)/|\mathbb{S}^1|$. Conversely, if $x_1, \ldots, x_n$ and*

$w_1, \ldots, w_n$ *form a GCF for* $\mathrm{Hom}_{2e}(\mathbb{R}^2)$ *with respect to* $\int_{\mathbb{S}^1} \cdot \rho(d\omega)/|\mathbb{S}^1|$, *then points* ω_i *and weights* $w_i \int_0^1 r^{2e+1} \, dr$ *form a cubature formula for* $\mathrm{Hom}_{2e}(\mathbb{R}^2)$ *with respect to* $\int_{\mathbb{B}^2} \cdot d\omega$.

Proof The result can be easily shown by observing that for all $f \in \mathrm{Hom}_{2e}(\mathbb{R}^2)$

$$\int_{\mathbb{B}^2} f(\omega) \, d\omega = \int_0^1 \left(\int_{\mathbb{S}^1} f(r\omega) \, \rho(d\omega) \right) r \, dr = \int_0^1 r^{2e+1} dr \int_{\mathbb{S}^1} f(\omega) \, \rho(d\omega). \qquad \square$$

Proof of Theorem 5.8. By Lemma 5.7, it suffices to consider a cubature formula for $\mathrm{Hom}_{2e}(\mathbb{R}^2)$ for integral $\int_{\mathbb{B}^2} \cdot d\omega$. For further arguments below, it is remarked that

$$\dim \mathrm{Hom}_e(\mathbb{R}^2) = e + 1,$$
$$\mathrm{Hom}_{2e}(\mathbb{R}^2) = \mathrm{Hom}_e(\mathbb{R}^2) \cdot \mathrm{Hom}_e(\mathbb{R}^2).$$

To $\omega = (\omega_1, \omega_2) \in B^2 \setminus \{0\}$ associate the complexification z_ω, that is,

$$z_\omega = r_\omega e^{\sqrt{-1}\theta_\omega} \in \mathbb{C}$$

with $r_\omega := |z_\omega| = \|\omega\|$ and $\theta_\omega := \mathrm{Arg}\, z_\omega$. Note that $\{z^i \bar{z}^{-i} \mid i = 0, \ldots, e\}$ is a basis of $\mathrm{Hom}_e(\mathbb{R}^2)$. Then the kernel function K for $\mathrm{Hom}_e(\mathbb{R}^2)$ is given by

$$K(\omega, \omega') = \frac{1}{\pi}(e+1)r_\omega^e r_{\omega'}^e \sum_{i=0}^e z_\omega^i \bar{z}_\omega^{e-i} z_{\omega'}^{l-i} \bar{z}_{\omega'}^{-i}$$

$$= \frac{1}{\pi} \sum_{l=e-1}^e (l+1)r_\omega^l r_{\omega'}^l \sum_{i=0}^l \exp(\sqrt{-1}(2i - l)(\theta_\omega - \theta_{\omega'}))$$

$$= \frac{1}{\pi}(e+1)r_\omega^e r_{\omega'}^e \sum_{i=0}^l \cos(2i - l)(\theta_\omega - \theta_{\omega'}).$$

Then it follows that

$$K(\omega, \omega') = \begin{cases} \dfrac{1}{\pi}(e+1)^2 r_\omega^e r_{\omega'}^e (-1)^{ek} & \text{if } \theta_\omega - \theta_{\omega'} = k\pi; \\[2mm] \dfrac{1}{\pi}(e+1)r_\omega^e r_{\omega'}^e \dfrac{\sin((e+1)(\theta_\omega - \theta_{\omega'}))}{\sin(\theta_\omega - \theta_{\omega'})} & \text{if } \theta_\omega - \theta_{\omega'} \notin \pi\mathbb{Z}. \end{cases} \qquad (5.11)$$

Now, let (X, w) be a cubature formula for $\mathrm{Hom}_{2e}(\mathbb{R}^2)$ with $e + 1$ points. Theorem 5.4 (i) and (5.11) show that

$$w_x = \frac{\pi}{(e+1)^2 \|x\|^{2e}} \quad \text{for every } x \in X, \tag{5.12}$$

$$\theta_x - \theta_{x'} \in \frac{\pi}{e+1} \mathbb{Z} \setminus \pi \mathbb{Z} \quad \text{for every distinct } x, x' \in X.$$

Hence, there exist nonzero $p_i \in [-1, 1]$ and $\theta \in [0, \pi/(e+1)]$ such that

$$\{z_x \mid x \in X\} = \{p_i e^{\sqrt{-1}(i\pi/(e+1))+\theta} \mid i = 0, \ldots, e\}. \tag{5.13}$$

Conversely, if X and w are of the form (5.13) and (5.12), respectively, then by (5.11) and Theorem 5.4 (ii), it is seen that (X, w) is a cubature formula for $\mathrm{Hom}_{2e}(\mathbb{R}^2)$ with $e + 1$ points. $\qquad\square$

Remark 5.3 (i) A function of type, $\sin((t+1)u)/\sin u$, appearing in (5.11) is widely known as the *Chebyshev polynomial of the second kind* and often denoted by $U_t(\cos u)$ [32]. With a change of variables, $x = \cos u$, function $U_t(\cos u)$ in variable u is simply converted to a polynomial in variable x. Such polynomials are then orthogonal with respect to density function $\pi \sqrt{1 - x^2}/2$ on closed interval $[-1, 1]$, which are not only of great importance in the theory of orthogonal polynomials but also provide various attractive subjects in design of experiments, e.g., in the study of *limiting optimal design measures* on $[-1, 1]$ for large-degree polynomial regressions; for details, see Chap. 2 of Huda [17].
(ii) Reznick's proof of Theorem 5.8 utilizes various algebraic techniques from the theory of binary forms; see [27, Chap. 5] for details. Lyubich and Vaserstein [19] have given an alternative proof of Theorem 5.8 by using some basic facts about spherical cubature, e.g., Hong's characterization [16] of tight spherical cubature which is based on the *Newton formula for power sums*.

One more implication of the basic results are looked at.

Let $\mathscr{H}$ be a space of L^2-integrable $\mathbb{K}$-valued functions on a set Ω. Let $f_1, \ldots, f_N$ be a basis of $\mathscr{H}$ with $N = \dim_{\mathbb{K}} \mathscr{H}$. Let $F = (f_1, \ldots, f_N)^T$. Start with a linear regression model

$$Y(\omega) = F(\omega)^T \theta + \varepsilon_\omega, \qquad \omega \in \Omega \tag{5.14}$$

where $\theta = (\theta_1, \ldots, \theta_N)^T$ is a vector of unknown parameters θ_i and ε_ω is a random error such that $\mathbf{E}[\varepsilon_\omega] = 0$ and $\mathbf{E}[\varepsilon_\omega \varepsilon_{\omega'}] = \sigma^2$ or 0 according as $\omega = \omega'$ or not. Clearly, this model generalizes the classical polynomial model (3.1).

A positive Borel measure μ on Ω, called a *design*, defines an inner product

$$(g, h)_{\mathscr{H}} = \int_\Omega g(\omega)\overline{h(\omega)}\, \mu(d\omega), \qquad g, h \in \mathscr{H}.$$

In order that θ be estimable, our attention is restricted to designs μ with positive definite information matrix $\mathbf{M}(\mu) = ((f_i, f_j)_{\mathscr{H}})_{1 \leq i, j \leq N}$.

As already seen in Chap. 3, a fundamental problem in design of experiments is to find a finite set of observation locations in Ω, to represent a design μ which is optimal with respect to some statistical criterion; see, e.g., [21, 24]. In a classical polynomial regression model,[3] this is just reduced to the existence problem on cubature of a given degree. Similarly for model (5.14), we will approximate a continuous design by a discrete design. As long as a space $\mathscr{H}$ of L^2-functions is taken into consideration, Theorem 5.2 ensures that such a discretization is always possible.

Theorem 5.9 ([29])   *Let μ be a (possibly infinite) design on a measurable space $(\Omega, \mathscr{F})$ concentrated in $A \in \mathscr{F}$, with information matrix $\mathbf{M}(\mu)$ on a finite-dimensional vector space $\mathscr{H}$ of L^2-functions on Ω. Then there exist a positive integer $n \leq \dim \mathscr{H} \cdot \mathscr{H}^- + 1$, $x_1, \ldots, x_n \in A$ and $w_1, \ldots, w_n > 0$ such that $\mathbf{M}(\mu) = \mathbf{M}(\xi)$ with $\xi = \sum_{i=1}^{n} w_i \delta_{x_i}$. In particular if μ is optimal, then so is ξ. Moreover, if $\mathscr{H}$ contains constant functions, then the evaluation of value n can be improved by one.*

Corollary 5.5 ([21, Theorem 5.4]).   *Let μ be a probability measure on a subset Ω of $\mathbb{R}^d$, with information matrix $\mathbf{M}(\mu)$ on $\mathscr{P}_e(\Omega)$. Then there exists a design ξ with $\mathbf{M}(\mu) = \mathbf{M}(\xi)$ having finite support of size at most $\dim \mathscr{P}_{2e}(\Omega)$. In particular, if μ is optimal, then so is ξ.*

Let us now look at how to estimate θ by making use of a Chebyshev-type cubature. Let μ be a (possibly infinite) design on a measurable space $(\Omega, \mathscr{F})$ with information matrix $\mathbf{M}(\mu)$ on $\mathscr{H}$. With samples $x_1, \ldots, x_n$, model (5.14) is reduced to

$$Y(x_i) = F(x_i)^T \theta$$

which can be written in the matrix form as

$$\mathbf{Y} = \mathbf{X}\theta + \varepsilon$$

where

$$\mathbf{Y} = \begin{pmatrix} Y(x_1) \\ \vdots \\ Y(x_n) \end{pmatrix}, \quad \mathbf{X} = \begin{pmatrix} f_1(x_1) \cdots f_N(x_1) \\ \vdots \qquad \vdots \\ f_1(x_n) \cdots f_N(x_n) \end{pmatrix}, \quad \theta = \begin{pmatrix} \theta_1 \\ \vdots \\ \theta_N \end{pmatrix}, \quad \varepsilon = \begin{pmatrix} \varepsilon_1 \\ \vdots \\ \varepsilon_n \end{pmatrix}.$$

As in (3.2), the least square theory shows that $\hat{\theta} = (\mathbf{X}^T\mathbf{X})^{-1}\mathbf{X}^T\mathbf{Y}$ and

$$\mathbf{E}[\hat{\theta}] = \theta, \qquad \mathrm{Cov}[\hat{\theta}] = \sigma^2(\mathbf{X}^T\mathbf{X})^{-1}.$$

If $x_1, \ldots, x_n$ are the points of a Chebyshev-type cubature for $\mathscr{H}$ (repeated points allowed), then $\mathrm{Cov}[\hat{\theta}]$ can be directly connected with information matrix $\mathbf{M}(\mu)$ as follows:

$$\mathrm{Cov}[\hat{\theta}] = \frac{\sigma^2}{n}\mathbf{M}^{-1}(\mu).$$

---

[3]Namely, $\mathscr{H}$ is a polynomial space. A design is said to have *degree e* if $\mathscr{H} = \mathscr{P}_{2e}(\Omega)$; see [21].

This fact follows since

$$\sigma^2(\mathbf{X}^T\mathbf{X})^{-1} = \sigma^2\left(\sum_{l=1}^n f_i(x_l)\overline{f_j(x_l)}\right)^{-1}_{i,j} = \frac{\sigma^2}{n}\left(\frac{1}{n}\sum_{l=1}^n f_i(x_l)\overline{f_j(x_l)}\right)^{-1}_{i,j}$$
$$= \frac{\sigma^2}{n}\left(\int_\Omega f_i(\omega)\overline{f_j(\omega)}\,\mu(d\omega)\right)^{-1}_{i,j}.$$

This is a full generalization of (3.4) in Remark 3.1.

By Theorem 5.3, the following result can be obtained.

Theorem 5.10 ([29]) *Let Ω be a path-connected topological space. Let μ be a (possibly infinite) design on Ω with full support, and with information matrix $\mathbf{M}(\mu)$ on a finite-dimensional vector space $\mathcal{H}$ of continuous L^2-functions. Then there exists a design ξ with $\mathbf{M}(\mu) = \mathbf{M}(\xi)$ and finite support X with rational weights. In particular, if μ is optimal, then so is ξ.*

In the course of the proof of Corollary 5.2, Seymour and Zaslavsky [30, Sect. 8] prove the existence of a cubature with rational weights, along with the evaluation of the number of points needed. For instance, if Ω is an analytic manifold and $\mathcal{H}$ consists of analytic functions, then the number of points in a rational-weighted cubature is bounded from above by $\dim\mathcal{H} + 1$. Moreover if the connectivity of Ω is assumed, then the number of points can be reduced by one [9]. Replacing the Carathéodory theorem by this improvement in the argument of [30, pp. 219–220] gives the following result (Theorem 5.11).

Theorem 5.11 *Let Ω be a connected analytic manifold. Let μ be a (possibly infinite) measure on Ω, with information matrix $\mathbf{M}(\mu)$ on a finite-dimensional vector space $\mathcal{H}$ of analytic L^2-functions. Then there exist a positive integer $n \leq \dim\mathcal{H}\cdot\mathcal{H}^-$, points $x_1, \ldots, x_n \in \Omega$ and positive rational numbers $w_1, \ldots, w_n$ such that $\mathbf{M}(\mu) = \mathbf{M}(\xi)$ with $\xi = \sum_{i=1}^n w_i\delta_{x_i}$. In particular, if μ is optimal, then so is ξ.*

This theorem, together with [21, Theorem 5.1], shows a strengthening of a theorem of Neumaier and Seidel [21, Corollary 5.5].

Corollary 5.6 *For any positive integers d and e, there exists an optimal experimental design of degree e on unit sphere $\mathbb{S}^{d-1}$, with at most $\dim\mathscr{P}_{2e}(\mathbb{S}^{d-1})$ points.*

For statistical applications, the number n of points in Theorem 5.11 will be desirable to be small rather than large and thus our results, Theorems 5.9 and 5.11, may be of no practical importance. But, we know that, in general regression model (5.14), any optimal design μ can be represented by a discrete design with rational weights, if μ is an infinite measure or a finite measure without full-support assumption.

Finally, we want to mention more. The authors do not know whether infinite designs find actual applications in statistical experiments, but Professor Takafumi Kanamori mentions the following implication of cubature for infinite measures: In Bayesian inference the posterior distribution $p(\omega'|\omega)q(\omega)/\int_\Omega p(\omega'|\omega)q(\omega)d\omega$ is used for statistical prediction problems. Here, $p(\omega'|\omega)$ is a statistical model of a quantity ω' with parameter $\omega \in \Omega$ and $q(\omega)$ is a prior distribution defined over parameter space Ω. Although a probability measure is often adopted as the prior distribution, there are some situations that we need to consider an infinite design $q(\omega)d\omega$, i.e., the Bayesian estimator with an improper prior. As far as L^1-functions $p(\omega'|\omega)$ (e.g., for finitely many data ω') are picked up, a cubature for $\int_\Omega q(\omega)d\omega$ can always be obtained by Theorem 5.2, which may have a practical impact for computation of the posterior defined from the improper prior. Then a challenging problem will be to present explicit constructions of cubature for general functional spaces for infinite measures.

References

1. Bachoc, C., Coulangeon, R., Nebe, G.: Designs in Grassmannian spaces and lattices. J. Algebr. Comb. **16**(1), 5–19 (2002)
2. Bannai, E., Hoggar, S.G.: On tight t-designs in compact symmetric spaces of rank one. Proc. Japan Acad. Ser. A Math. Sci. **61**(3), 78–82 (1985)
3. Bayer, C., Teichmann, J.: The proof of Tchakaloff's theorem. Proc. Am. Math. Soc. **134**(10), 3035–3040 (2006)
4. Box, G.E.P., Behnken, D.: Some new three level designs for the study of quantitative variables. Technometrics **2**(4), 455–475 (1960)
5. Box, G.E.P., Hunter, J.S.: Multi-factor experimental designs for exploring response surfaces. Ann. Math. Stat. **28**(1), 195–241 (1957)
6. Box, G.E.P., Hunter, J.S., Hunter, W.G.: Statistics for Experimenters, 2nd edn. Wiley Series in Probability and Statistics. Wiley-Interscienced, Hoboken, NJ (2005)
7. Cox, D.A., Little, J., O'Shea, D.: Ideals, Varieties, and Algorithms. Undergraduate Texts in Mathematics, 4th edn. Springer, Cham (2015)
8. Curto, R.E., Fialkow, L.A.: A duality proof of Tchakaloff's theorem. J. Math. Anal. Appl. **269**(2), 519–532 (2002)
9. Danzer, L., Grünbaum, B., Klee Jr., V.L.: Helly's theorem and its relatives. In: Proceedings of Symposia Pure Mathematics, vol. 7, pp. 101–180. American Mathematical Society, Providence, RI (1963)
10. Delsarte, P., Goethals, J.M., Seidel, J.J.: Spherical codes and designs. Geom. Dedicata **6**(3), 363–388 (1977)
11. Farrell, R.H., Kiefer, J., Walbran, A.: Optimum multivariate designs. In: Proceedings of Fifth Berkeley Sympos., vol. 1, pp. 113–138. University of California Press, Berkeley, California (1967)
12. Fisher, R.A.: An examination of the different possible solutions of a problem in incomplete blocks. Ann. Eugen. **10**, 52–75 (1940)
13. de la Harpe, P., Pache, C.: Cubature formulas, geometrical designs, reproducing kernels, and Markov operators. In: Infinite Groups: Geometric. Combinatorial and Dynamical Aspects. Progress in Mathematics, vol. 248, pp. 219–267. Birkhäuser, Basel (2005)
14. Hirao, M., Okuda, T., Sawa, M.: Some remarks on cubature formulas with linear operators. J. Math. Soc. Jpn. **68**(2), 711–735 (2016)
15. Hirao, M., Sawa, M., Jimbo, M.: Constructions of Φ_p-optimal rotatable designs on the ball. Sankhyā Ser. A **77**(1), 211–236 (2015)
16. Hong, Y.: On spherical t-designs in R^2. Eur. J. Comb. **3**(3), 255–258 (1982)
17. Huda, S.: Rotatable designs: constructions and considerations in the robust design of experiments. Ph.D. thesis, Imperial College, University of London (1981)

18. Ionin, Y.J., van Trung, T.: Symmetric designs. In: Handbook of Combinatorial Designs, Second edn., pp. 110–124. CRC Press, Boca Raton, USA (2007)
19. Lyubich, Y.I., Vaserstein, L.N.: Isometric embeddings between classical Banach spaces, cubature formulas, and spherical designs. Geom. Dedicata **47**(3), 327–362 (1993)
20. Meyer, B.: Extreme lattices and vexillar designs. J. Algebra **322**(12), 4368–4381 (2009)
21. Neumaier, A., Seidel, J.J.: Measures of strength $2e$ and optimal designs of degree e. Sankhyā Ser. A **54**, 299–309 (1992)
22. Nozaki, H., Sawa, M.: Remarks on Hilbert identities, isometric embeddings, and invariant cubature. Algebra i Analiz **25**(4), 139–181 (2013)
23. Petrenjuk, A.Y.: Fisher's inequality for tactical configurations. Mat. Zametki **4**, 417–424 (1968)
24. Pukelsheim, F.: Optimal Design of Experiments. Classics in Applied Mathematic, vol. 50. Society for Industrial and Applied Mathematics (SIAM), Philadelphia, PA (2006). Reprint of the 1993 original
25. Putinar, M.: A note on Tchakaloff's theorem. Proc. Am. Math. Soc. **125**(8), 2409–2414 (1997)
26. Ray-Chaudhuri, D.K., Wilson, R.M.: On t-designs. Osaka J. Math. **12**(3), 737–744 (1975)
27. Reznick, B.: Sums of even powers of real linear forms. Mem. Am. Math. Soc. **96**(463) (1992)
28. Roy, A.: Bounds for codes and designs in complex subspaces. J. Algebr. Comb. **31**, 1–32 (2010)
29. Sawa, M.: The theory of cubature formulas (in Japanese). Sugaku **68**(1), 24–53 (2016)
30. Seymour, P.D., Zaslavsky, T.: Averaging sets: a generalization of mean values and spherical designs. Adv. Math. **52**(3), 213–240 (1984)
31. Shamsiev, E.A.: Cubature formulas for a disk that are invariant with respect to groups of transformations of regular polyhedra into themselves. Comput. Math. Math. Phys. **46**(7), 1147–1154 (2006)
32. Szegő, G.: Orthogonal Polynomials. Colloquium Publications, vol. XXIII. American Mathematical Society, Providence, R.I (1975)
33. Victoir, N.: Asymmetric cubature formulae with few points in high dimension for symmetric measures. SIAM J. Numer. Anal. **42**(1), 209–227 (2004)
34. Xu, Y.: Lower bound for the number of nodes of cubature formulae on the unit ball. J. Complexity **19**(3), 392–402 (2003)

Correction to: Euclidean Design Theory

Correction to:
M. Sawa et al., *Euclidean Design Theory*,
JSS Research Series in Statistics,
https://doi.org/10.1007/978-981-13-8075-4

In the original version of the book, the following belated corrections have been made.

In the following list, line n means n-th line from the top and line $-n$ means n-th line from the bottom:

Page 4, line -7: Proof of Proposition 1.1(ii),

$$ \text{``} \sum_{i,j} c_i c_j K_{ij}^{(1)} K_{ij}^{(2)} = \sum_{i,j} c_i c_j \left(\sum_k \lambda_k v_{ki} v_{kj} \right) \sum_{i,j} K_{ij}^{(2)} \text{''} $$

has been replaced by

$$ \text{``} \sum_{i,j} c_i c_j K_{ij}^{(1)} K_{ij}^{(2)} = \sum_{i,j} c_i c_j \left(\sum_k \lambda_k v_{ki} v_{kj} \right) K_{ij}^{(2)} \text{''}. $$

Page 21, line 1: "$\int_{-1}^{1} \cdot \, du/2$" has been replaced by "$\int_0^1 \cdot \, du$".

The updated versions of these chapters can be found at
https://doi.org/10.1007/978-981-13-8075-4_1
https://doi.org/10.1007/978-981-13-8075-4_2
https://doi.org/10.1007/978-981-13-8075-4_4

Page 80, line -7 has been changed as follows:

$$\frac{1}{(1-u^4)(1-u^6)\cdots(1-u^{2d-2})(1-u^d)} = \begin{cases} 1+2u^4+u^6+3u^8+\mathcal{O}(u^{10}), & \text{if} \quad d=4, \\ 1+u^4+u^5+u^6+2u^8+\mathcal{O}(u^9), & \text{if} \quad d=5, \\ 1+u^4+2u^6+2u^8+\mathcal{O}(u^{10}), & \text{if} \quad d=6, \\ 1+u^4+u^6+u^7+2u^8+\mathcal{O}(u^{10}), & \text{if} \quad d=7, \\ 1+u^4+u^6+3u^8+\mathcal{O}(u^{10}), & \text{if} \quad d=8, \\ 1+u^4+u^6+2u^8+\mathcal{O}(u^{10}), & \text{if} \quad d\geq9. \end{cases}$$

Page 81, line 7 has been changed as follows:

$$\dim \mathrm{Harm}_8 \left(\mathbb{R}^d\right)^{D_d} = \begin{cases} 3, & d=4,8 \\ 2, & d\geq5, d\neq8. \end{cases}$$

Page 81, line -10: "$d\geq4, d\neq6$" has been replaced by "$d\geq4$".

Page 81, line -6: "the following polynomials $f_{8,1}, f_{8,2}, f_{8,3}$, respectively, as" has been replaced by "the following polynomials $f_{8,1}, f_{8,2}, f_{8,3}, f_{8,4}$, respectively, as".

Page 82, line 1: the following has been included:
For $d=8$

$$f_{8,4} = x_1 x_2 x_3 x_4 x_5 x_6 x_7 x_8.$$

Page 82, line -10: "For $d\geq4, d\neq6$" has been replaced by "For $d\geq4$".

Page 83, line 5: the following has been included:

(iv) For $d=8$

$$f_{8,4}(v_1) = \cdots = f_{8,4}(v_6) = 0, \quad f_{8,4}(v_7) = -\frac{1}{4096}, \quad f_{8,4}(v_8) = \frac{1}{4096}.$$

Page 99, line 8 has been changed as follows:

$$f_{8,2}(v_{A,s}) = \frac{1}{(A+s)^4}\left\{A^2 s + \frac{(s-1)s}{2} - \frac{6}{d-2}\left(A(s-1)s + \frac{A^2(s-1)s}{2}\right.\right.$$
$$\left.\left. + \frac{(s-2)(s-1)s}{2}\right) + \frac{9s(-3+4A+s)(s-2)(s-1)}{2(d-3)(d-2)}\right\}.$$

The correction chapters have been updated with these changes.

Index

A
Addition formula, 9, 12, 113
Adjoint, 114
Algebraic Code
 BCH code, 98
 Delsarte–Goethals code, 98
Approximate design, 48, 126
Aronszajn theorem, 2

B
Banach space, 40
Bayesian inference, 127
Binary hypercube, 106
Binary sphere, 105
Block design
 t-wise balanced design, 94, 104
 balanced incomplete block (BIB) design, 94, 106, 115
 combinatorial t-design, 31, 94, 115
 doubly BIB design, 94
Bochner theorem, 16
Bondarenko-Radchenko-Viazovska theorem, 33
Bound
 Fisher-type bound, 14, 19, 24, 38, 54
 generalized linear programming (LP) bound, 117
 linear programming (LP) bound, 11, 38, 116
 Möller bound, 19, 25, 38, 121
 Stroud bound, 19, 24, 38, 54
Boundedness of linear operator, 7

C
Carathéodory theorem, 23, 127

Cauchy–Schwarz inequality, 117
Centrally symmetric configuration, 94
Central symmetry of integral, 24
Characteristic function, 31, 95, 104, 105
Christoffel–Darboux formula, 6, 8
Christoffel number, 6
Classical orthogonal polynomial, 13
 Chebyshev polynomial of the first kind, 32
 Chebyshev polynomial of the second kind, 125
 Gegenbauer polynomial, 9, 38
 Hermite polynomial, 6, 8
 Jacobi polynomial, 11, 13
 scaled Gegenbauer polynomial, 10, 12, 113
Compact formula, 8, 10
Continuous design, 48, 126
Convexity, 23, 110, 117
Corner vector, 63, 65, 67, 71, 83
Corner vector method, 54, 63, 83, 98
Coxeter element, 65
Coxeter graph, 65
Cross polytope, 34
Cubature formula, 19
 Chebyshev-type, 31
 degree-type, 20, 104
 index-type, 104, 116
 Laplacian-type, 122
 minimal cubature, 27, 33
 rational-weighted cubature, 127

D
Degree
 of cubature formula, 20
 of experimental design, 47, 126

© The Author(s), under exclusive license to Springer Nature Singapore Pte Ltd. 2019
M. Sawa et al., *Euclidean Design Theory*, JSS Research Series in Statistics,
https://doi.org/10.1007/978-981-13-8075-4

Demihypercube, 83
Design matrix, 46, 94
Determinantal point process, 40
Dihedral group, 65, 74
Dimension of functional space, 11, 26
Discriminant analysis, 5
Double-counting technique, 105

E

Equiangular line, 34
Euclidean design, 15, 19, 34
    degree-type, 34, 73, 84
    D-optimality, 45, 53, 55, 84, 99, 107
    index-type, 40, 104
    tight design, 36, 116
Euclidean design theory, 103
Experimental design, 47, 125
Exponent, 65, 67, 69, 71, 75

F

Feature map, 16, 98
    random Fourier feature, 16
Fischer decomposition, 13, 35, 37, 118
Fisher inequality, 108, 115
Fourier expansion, 7
Fractional factorial design, 103
    Box–Behnken design, 31, 83, 94, 97, 107
    Box–Hunter polygonal design, 68, 70, 83
    central composite design, 70, 72, 108
    half-fractional factorial design, 94
    Plackett–Burman design, 108
    Scheffé lattice design, 68, 83
Fréchet derivative, 59
Fundamental root, 66, 68, 71
Fundamental system, 64

G

Gamma function, 11, 32
Gâteaux derivative, 59
Gauss–Markov theorem, 46
Gauss mean-value property, 122
Generalized cubature formula (GCF), 104
    Chebyshev-type, 104
    tight design, 121
Generalized Fisher inequality, 113
Generalized Sobolev theorem, 74, 84 120
Generalized Tchakaloff theorem, 109
Gram matrix, 3, 47
Gram–Schmidt orthonormalization, 6
Group orbit, 67, 69, 71

H

Harmonic Molien-Poincaré series, 75, 76, 79, 80
Harmonic polynomial, 9
Hermite interpolation, 122
Hilbert identity, 41, 123
    rational identity, 41
Hilbert–Schmidt operator, 12
Hilbert space, 2
    reproducing kernel Hilbert space (RKHS), 2, 39
Homogeneous polynomial, 9
Hyperoctahedral group, 108
Hyperoctahedron, 83

I

Icosahedron, 34
Infinite design, 127
Information matrix, 3, 46, 47, 53, 56, 125
Inner product
    Euclidean inner product, 4, 20
    L^2-inner product, 3, 47, 112
    scalar (inner) product, 114
Invariance of function, 73, 76, 79, 81, 119
Invariance of measure
    central symmetry, 7
    $\mathcal{O}(d)$-invariance, 48, 51, 56
    shift-invariance, 16
Invariance of measure space, 119
Invariance of weighted pair, 36, 73, 84
Irreducibility of reflection group, 65

J

Jittered sampling, 40

K

Kernel function (reproducing kernel), 1, 23, 50, 112, 113, 124
    Gaussian kernel, 16
    Hilbert–Schmidt kernel, 12
    kernel polynomial, 3, 6, 22, 28, 113
    modified kernel polynomial, 5, 29
    reproducing property, 2
    shift-invariant kernel, 16
Kiefer characterization theorem, 52, 56, 66
Kiefer general equivalence theorem, 59
Kiefer–Wolfowitz equivalence theorem, 45, 50

L

Leech lattice, 34

Limiting optimal design, 125
Linear functional, 7, 25
Linear isometry of normed vector spaces, 32, 40, 114, 123
L^1-integrable function, 104
L^2-integrable function, 104

M

McLaughlin graph, 34
Measure
    Borel measure, 108
    counting measure, 115
    Dirac measure, 47
    full-support measure, 110
    product measure, 21
    push-forward measure, 109
    surface measure, 10, 22, 34, 113
    uniform measure on sphere, 104, 121
Missing configuration, 34
Multivariate Jacobi weight function, 10, 30, 39, 119
Mysovskikh theorem, 6, 29, 39

N

Neumaier–Seidel equivalence theorem, 35, 74
Newton formula for power sum, 125

O

Octahedron, 72, 107
Optimal experimental design, 21, 32, 107
Optimality criterion
    A-optimality, 49
    Cohn–Kumar universal optimality, 60
    D-optimality, 45, 49
    E-optimality, 49
    G-optimality, 3, 49
    maximin-distance optimality, 60
    minimax-distance optimality, 60
    ϕ-optimality, 58
Orthogonal array (OA), 95, 105
    BCH OA, 98
    Delsarte–Goethals OA, 98
Orthonormal basis (ONB), 3, 6, 48
Orthonormal system (ONS), 6

P

Parseval identity, 7
Polar coordinate system, 32
Potential energy, 14, 60

discrete Riesz potential, 60
f-potential, 60
Product rule, 17, 21

Q

Quadrature formula, 20, 37, 51
    Chebyshev–Gauss quadrature, 30, 31
    Gaussian quadrature, 30, 38
    Hermite–Gauss quadrature, 6, 21, 30
    Jacobi–Gauss quadrature, 21, 30
    Laguerre–Gauss quadrature, 21, 30
Quasi-Monte Carlo (QMC) method, 39

R

Reflection, 64, 67, 69, 71
Reflection group, 36, 64, 65, 84, 99
Regularity of t-wise balanced design, 95, 104
Regular polygon, 27, 33, 47, 68, 70, 74, 122, 123
Reynolds operator, 74, 119
Riesz representation theorem, 7, 12, 113
Root system, 34, 64
Rotatable design, 48, 51, 107

S

Seidel'nikov inequality, 15
Seymour–Zaslavsky theorem, 33, 111
Simplex, 33, 65, 68, 83, 106
Simpson rule, 20
Singular value decomposition (SVD), 4
Sobolev theorem, 73, 119
Space-filling design, 39
Spherical design, 15, 31, 116
    tight design, 33, 47, 74
Spherical symmetry of integral, 25, 35
    circularly (radially) symmetric integral, 35
    Gaussian integration, 8, 21, 24, 35, 39, 97
    spherical integration, 9, 10, 15, 31, 73, 108, 113
Support of measure, 23
Symmetric polynomial, 75
Symmetric region
    ball, 51
    concentric spheres, 26, 34
    sphere, 9, 22, 34
Symmetry group
    of demihypercube, 36, 65, 71, 80, 84, 91

of hyperoctahedron, 65, 68, 79, 84, 87, 108
of simplex, 65, 76, 84, 85

T
Tchakaloff theorem, 23, 54, 108
Thinning method, 63, 95, 107
    Kuperberg method, 23
    Smolyak method, 17, 23
    Victoir method, 23

V
Variance function, 48, 52, 107

Volume of confidence ellipsoid, 49

W
Waring problem, 41
Weighted summation rule, 20
Wilson–Petrenjuk inequality, 115

X
Xu compact formula, 29, 39, 119

Z
Zernike-type polynomial basis, 55

Printed in the United States
By Bookmasters